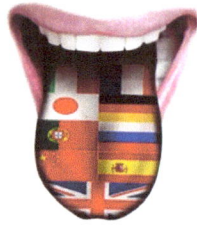

Editor's Comment

Global Energy Outlook - At A Glance in 24 Charts

Economic activity and population drive increases in energy use, global energy outlook projects that Non-OECD nations will drive the increase in total energy use, Non-OECD Asia accounts for 55 per cent of the world increase in energy use.

In this edition, you will be able to see at a glance the global energy outlook- In 24 interactive charts the industry's outlook in Liquid fuels markets, Natural gas markets, Electricity markets, and Energy-related carbon dioxide emissions.

You will read about the US Coal Mine basin and their total production outputs. Propane Exports Drove U.S. Petroleum Product Export Growth in First Half Of 2016.The impact of low oil prices on Russian's oil companies and government revenues. The largest power plants in the world and lots more.

The international pipeline, oil and gas safety conference March 14-16, 2017, seeks to address process safety issues in the upstream, midstream and downstream subsectors of the industry; with special focus on process safety, pipeline safety, and new regulatory impact. Early registration ends Dec. 28, 2016.

Conference agenda out! – get a copy @ http://oilandgassafetyconference.com

- Gloria Towolawi

USA Oil and Gas Monitor
A RGT Media Communications Corp.

Editor-in-Chief
Gloria Towolawi

Europe Bureau
Esther Coker

Nigeria Bureau
David Arhavbarien

Contributing Editor
Gloria Instead

Reporter
Caleb Motinwo

Advert & Marketing
Jewel Spring
T: 832-486-0095
E: advertise@usaoilandgasmonitor.com

Distribution & sales
Richard Godfirst

Subscribers Service
E: subscribe@usaoilandgasmonitor.com

RGT Media Communications Corp.
Publishers of
USA Oil and Gas Monitor
Workplace Weekly News
GlobalPRPlus

USA Oil and Gas Monitor is published 12 times a year monthly by RGT Media Communications Corp. 10777 Westheimer #1100

Houston, Texas 77042
Subscription price is $144 per year.
Digital copy $9.99 per download.

Copyright 2016 by
RGT Media Communications Corp.

Contents

Global Energy Outlook - At A Glance in 24 charts 4

Electric Power, Transportation and Industrial Sectors Largest Consumers of Energy Products 8

MECS Survey Preliminary estimates show that U.S. manufacturing energy consumption increased 3.7 percent between 2010 and 2014 11

U.S. Crude Oil Imports Increase During First Half Of 2016, The First Increase Since 2010 13

Three Gorges Dam, China Ranks First Out Of The Ten World's Largest Power Plants 15

Propane Exports Drove U.S. Petroleum Product Export Growth in First Half Of 2016 20

Russian Petroleum Companies and Government Revenues Declines Sharply Due to Low Oil Prices 22

BOEM's Draft Programmatic Environmental Impact Statement for Gulf of Mexico Geological and Geophysical Surveys Announced 24

PEIS Draft Regulations- API's Andy Radford Raise Concerns over Mitigation Measures Proposed 26

U.S. Total Coal Mine Production fell from between 2010 and 2014 27

The Energy Transition Could Come Faster Than We Think- Spencer Dale, BP's Group Chief Economist 28

Shell Divests Non-Core Shale Acreage in Western Canada For Total Consideration Of $1 Billion 30

API-Implementing EPA's Control Techniques Guidelines without proper scientific input is bad public policy 31

November 2016 • Issue 11

Renewables grow fastest, coal use plateaus, natural gas surpasses coal by 2030, and oil maintains its leading share
world energy consumption
quadrillion Btu

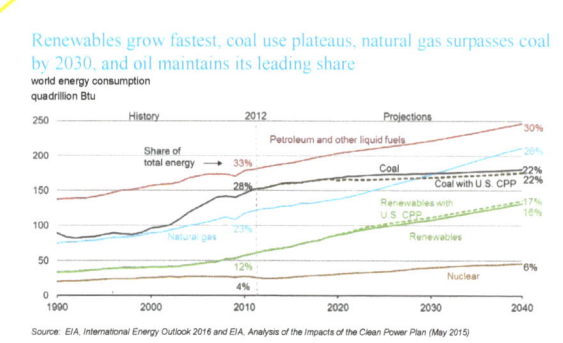

Source: EIA, International Energy Outlook 2016 and EIA, Analysis of the Impacts of the Clean Power Plan (May 2015)

As total energy consumption grows, shares by end-use sector remain relatively unchanged
world delivered energy consumption by end-use sector
quadrillion Btu

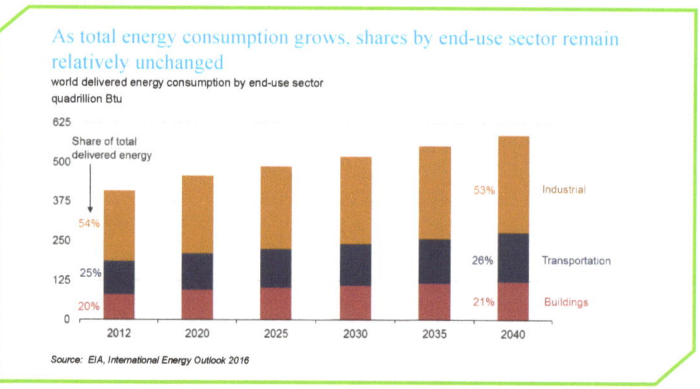

Source: EIA, International Energy Outlook 2016

Economic growth drives electricity demand; electricity use grows at a faster rate than other delivered energy, but slower than GDP
world GDP and net electricity generation
percent growth (rolling average of 3-year periods)

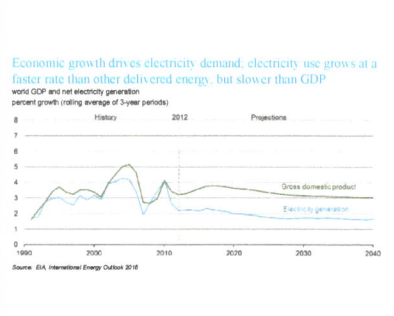

Source: EIA, International Energy Outlook 2016

Global Energy Outlook - At A Glance in 24 charts

Economic activity and population drive increases in energy use; energy intensity (E/GDP) improvements moderate this trend
average annual percent change (2012–40)
percent per year

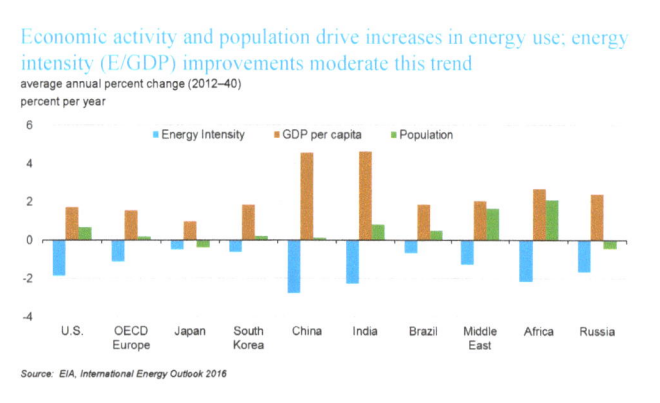

Source: EIA, International Energy Outlook 2016

Non-OECD nations drive the increase in total energy use
world energy consumption
quadrillion Btu

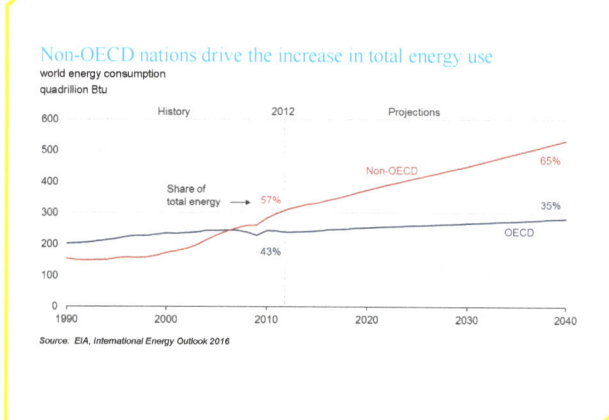

Source: EIA, International Energy Outlook 2016

Non-OECD Asia accounts for 55% of the world increase in energy use
world energy consumption
quadrillion Btu

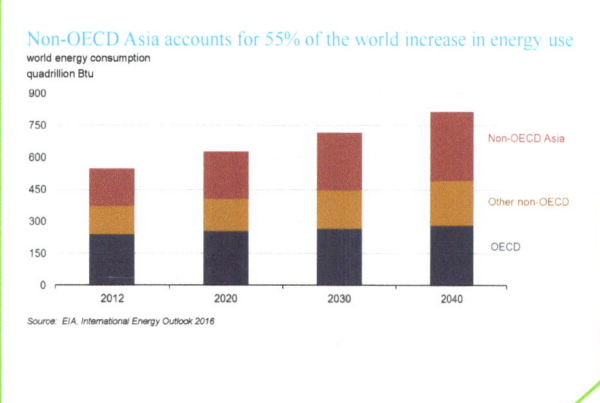

Source: EIA, International Energy Outlook 2016

Projected carbon intensity of energy use (CO2/E) declines through 2040 in both OECD and non-OECD; non-OECD CO2/E rose over 2000–12
carbon intensity of energy consumption, 1990-2040
kilograms CO2 per million Btu

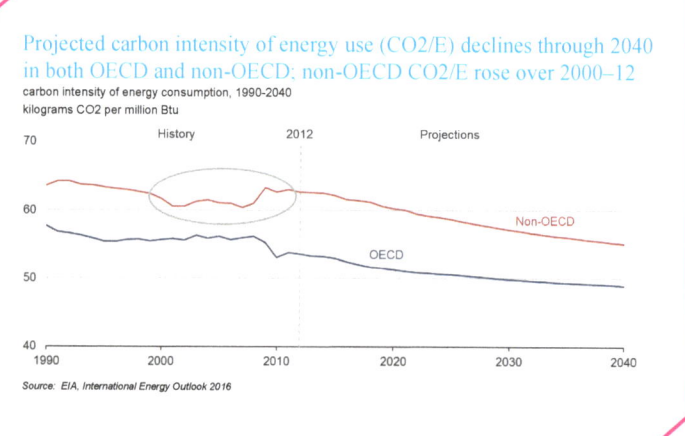

Source: EIA, International Energy Outlook 2016

Liquid Fuels Markets Outlook

Most of the growth in world oil consumption occurs in the non-OECD regions — especially Asia
world petroleum and other liquid fuels consumption
million barrels per day

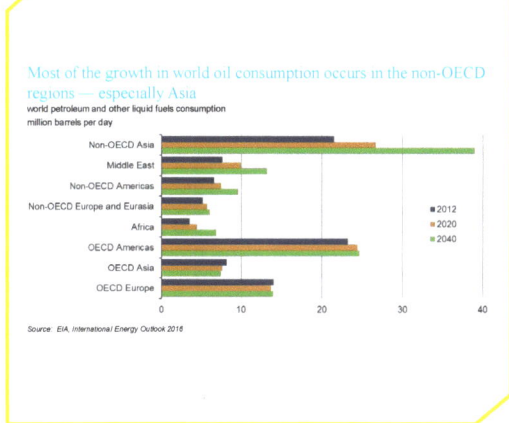

Source: EIA, International Energy Outlook 2016

Passenger-miles per person will rise as GDP per capita grows; travel growth is largely outside the OECD
passenger-miles per capita (left-axis) and GDP per capita (horizontal-axis) for selected country groupings 2010–40

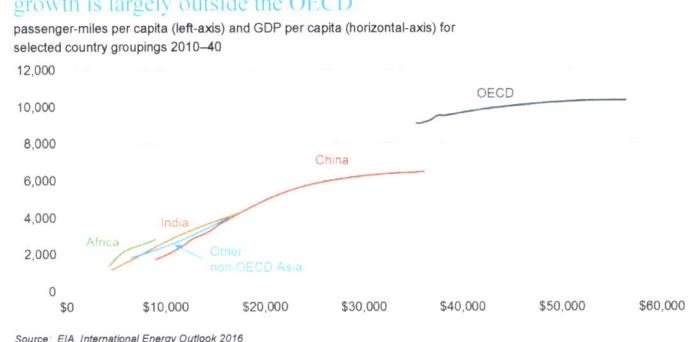

Source: EIA, International Energy Outlook 2016

Liquid fuels supplies from both OPEC and non-OPEC producers increase through 2040
world production of petroleum and other liquid fuels
million barrels per day

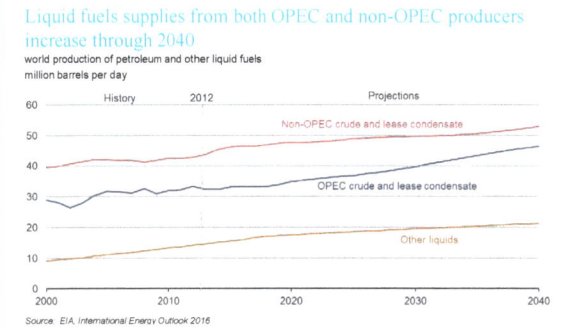

Source: EIA, International Energy Outlook 2016

Growth in OPEC production comes mainly from the Middle East
OPEC crude and lease condensate production
million barrels per day

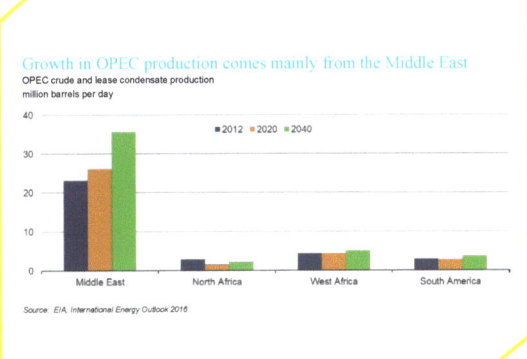

Source: EIA, International Energy Outlook 2016

Increases to non-OPEC oil supplies outside the United States are primarily from Brazil, Russia, Canada, and Kazakhstan
non-OPEC crude and lease condensate production in selected country groupings
million barrels per day

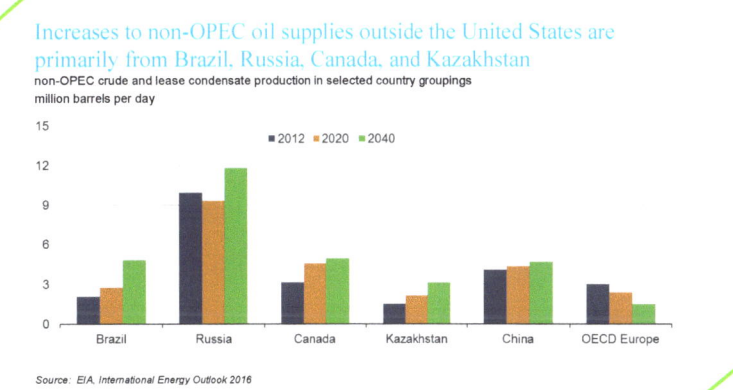

Source: EIA, International Energy Outlook 2016

The largest components of other liquid fuels are NGPL, refinery gain, and biofuels

million barrels per day

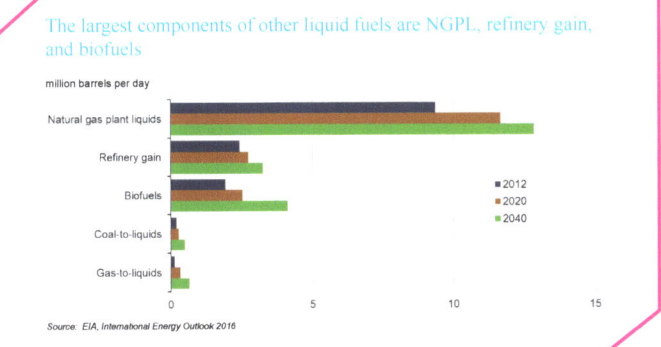

Source: EIA, International Energy Outlook 2016

November 2016 • Issue 11

Natural Gas Markets Outlook

Liquefaction capacity additions over the 2015-19 time period will increase global capacity by over 30%

LNG capacity additions
billion cubic feet per day

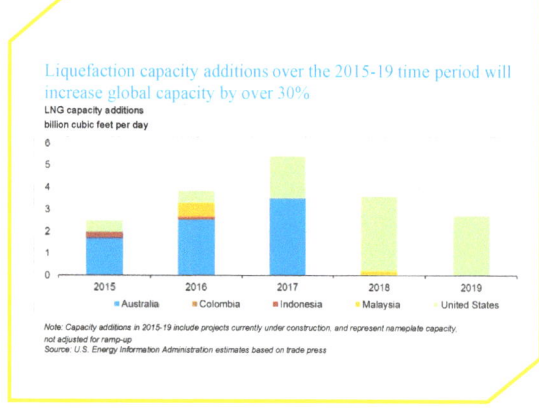

Note: Capacity additions in 2015-19 include projects currently under construction, and represent nameplate capacity, not adjusted for ramp-up
Source: U.S. Energy Information Administration estimates based on trade press

Shale gas, tight gas, and coalbed methane will become increasingly important to gas supplies, not only for the U.S., but also China and Canada

natural gas production by type
trillion cubic feet

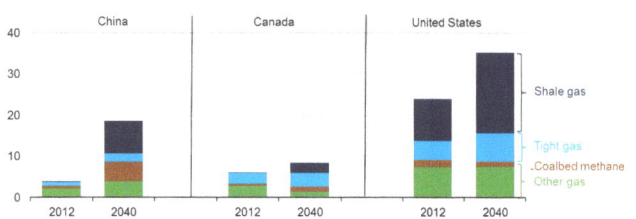

Note: Other natural gas includes natural gas produced from structural and stratigraphic traps (e.g. reservoirs), historically referred to as 'conventional' production.
Source: EIA, International Energy Outlook 2016

Non-OECD nations will account for 76% of the growth in natural gas consumption

world natural gas consumption
trillion cubic feet

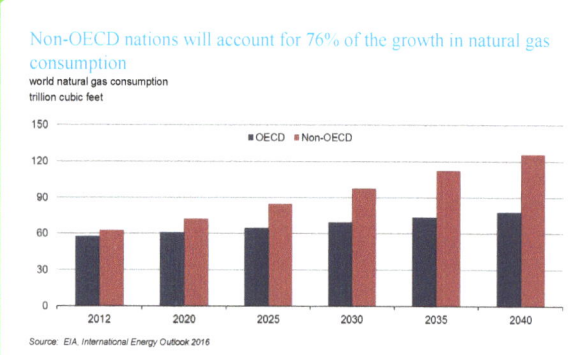

Source: EIA, International Energy Outlook 2016

Non-OECD Asia, Middle East, and OECD Americas account for the largest increases in natural gas production

world change in natural gas production, 2012–40
trillion cubic feet

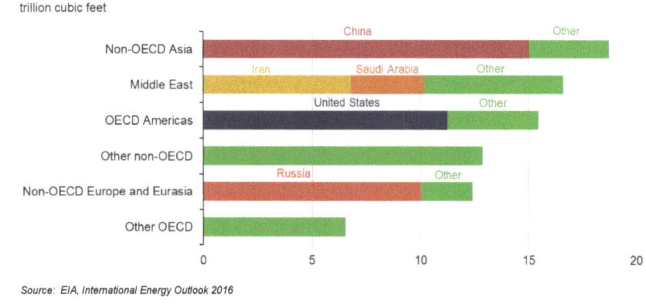

Source: EIA, International Energy Outlook 2016

Renewables, natural gas, and coal all contribute roughly the same amount of global net electricity generation in 2040

world net electricity generation by source
trillion kilowatthours

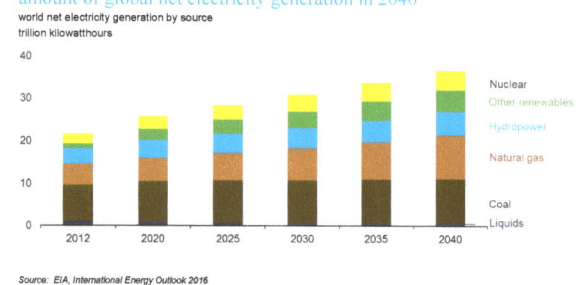

Source: EIA, International Energy Outlook 2016

Electricity Markets Outlook

GDP drives electricity demand growth, but the electricity growth rate compared to the GDP growth rate becomes smaller over time

world GDP and net electricity generation
percent growth (rolling average of 3-year periods)

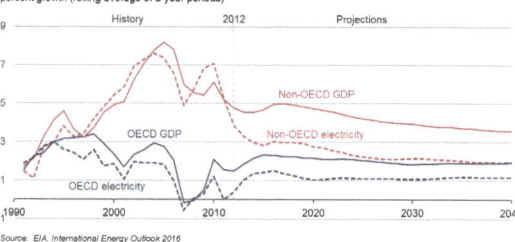

Source: EIA, International Energy Outlook 2016

Geographically, the scale and fuel mix of renewable generation in 2040 varies widely
renewable net electricity generation by selected country and country grouping, 2040
billion kilowatthours

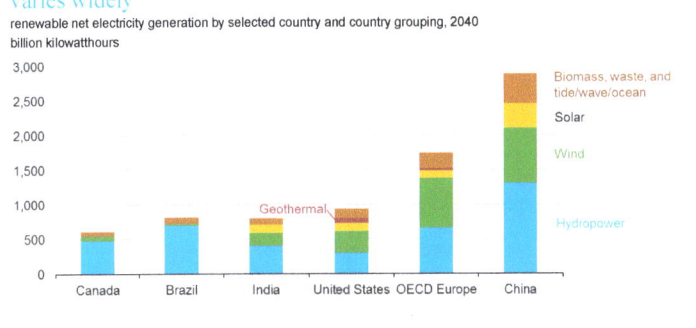

Source: EIA, International Energy Outlook 2016

Virtually all of the growth in nuclear power will occur in the non-OECD regions; China accounts for 61% of world nuclear capacity growth
world installed nuclear capacity by region
gigawatts

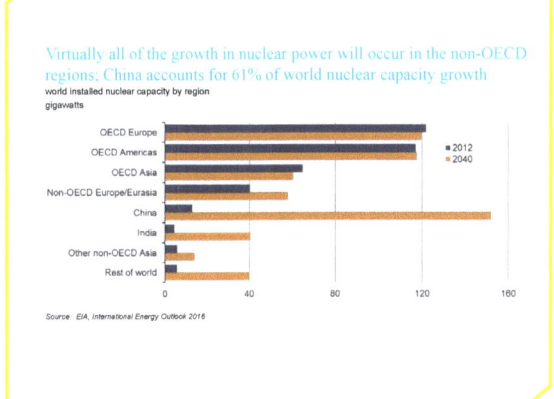

Source: EIA, International Energy Outlook 2016

Energy-related Carbon Dioxide Emissions Outlook

Coal remains the world's largest source of energy-related CO_2 emissions, but by 2040 its share declines to 38%
world energy-related carbon dioxide emissions
billion metric tons

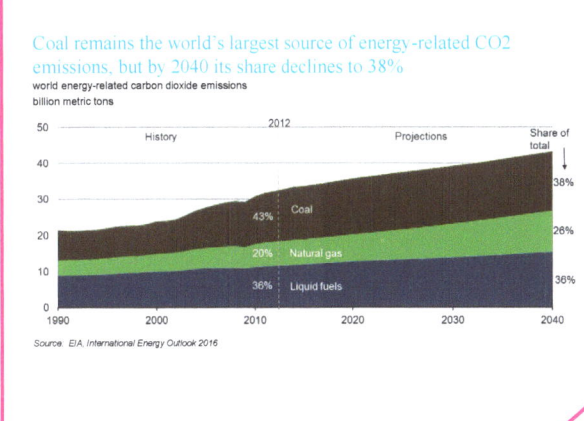

Source: EIA, International Energy Outlook 2016

Of the world's three largest coal consumers, only India is projected to continue to increase throughout the projection
coal consumption in the US, China, and India
quadrillion Btu

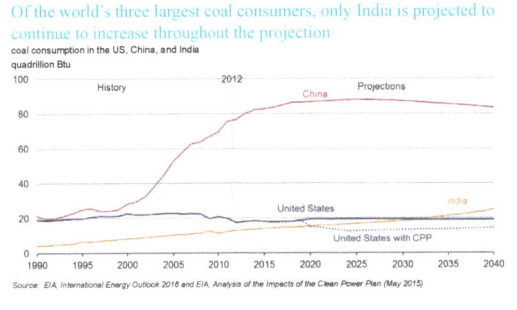

Source: EIA, International Energy Outlook 2016 and EIA, Analysis of the Impacts of the Clean Power Plan (May 2015)

Non-OECD Asia will account for about 60% of the world increase in energy-related CO_2 emissions
world energy-related carbon dioxide emissions
billion metric tons

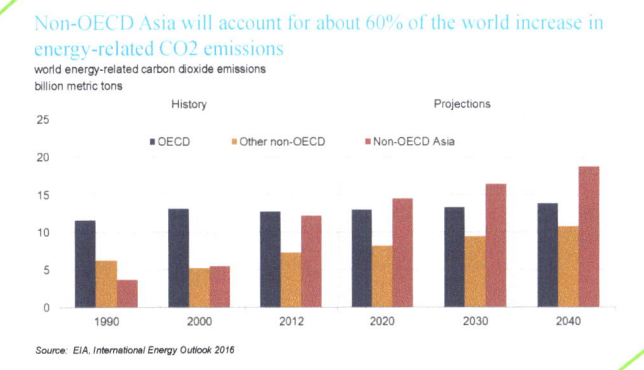

Source: EIA, International Energy Outlook 2016

November 2016 • Issue 11

U.S. primary energy consumption by source and sector, 2015
Total = 97.7 quadrillion British thermal units Btu

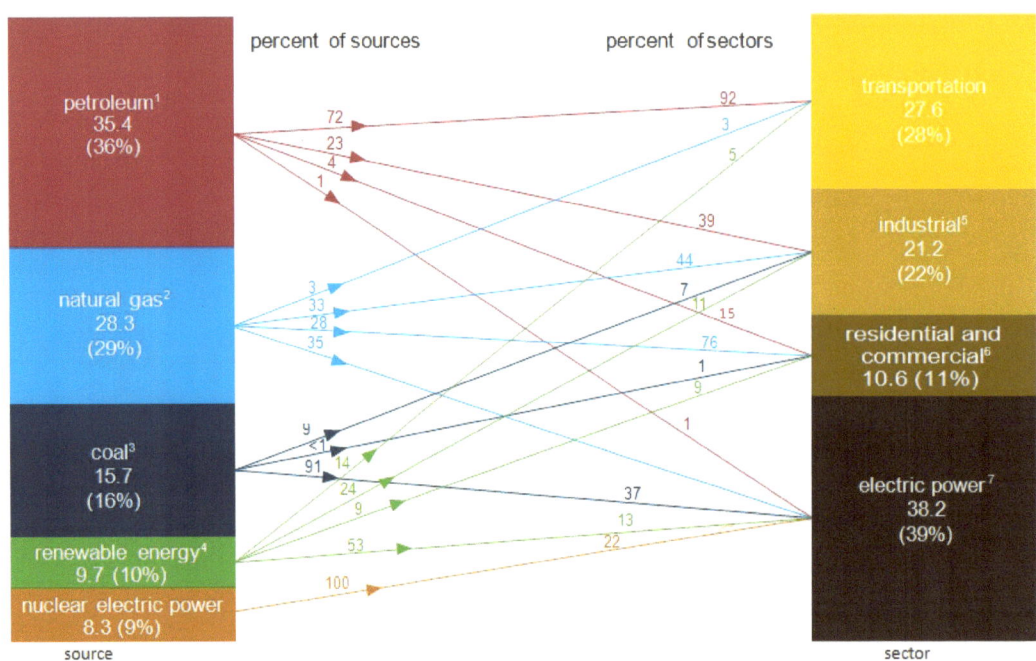

Electric Power, Transportation and Industrial Sectors Largest Consumers of Energy Products

U.S. energy-related carbon dioxide CO2 emissions totaled 2,530 million metric tons in the first six months of 2016. This was the lowest emissions level for the first six months of the year since 1991, as mild weather and changes in the fuels used to generate electricity contributed to the decline in energy-related emissions. EIA's Short-Term Energy Outlook projects that energy-associated CO2 emissions will fall to 5,179 million metric tons in 2016, the lowest annual level since 1992.

Mild weather. In the first six months of 2016, the United States had the fewest heating degree days an indicator of heating demand since at least 1949, the earliest year for which EIA has monthly data for all 50 states. Warmer weather during winter months reduces demand for heating fuels such as natural gas, distillate heating oil, and electricity. Overall, total primary energy consumption was 2 per cent lower compared with the first six months of 2015. The decrease was most notable in the residential and electric power sectors, where primary energy consumption decreased 9 per

Energy-related CO2 emissions for first half of 2016 lowest since 1991

Energy-related carbon dioxide emissions by source (Jan 1990 - Jun 2016)
12-month moving total, billion metric tons

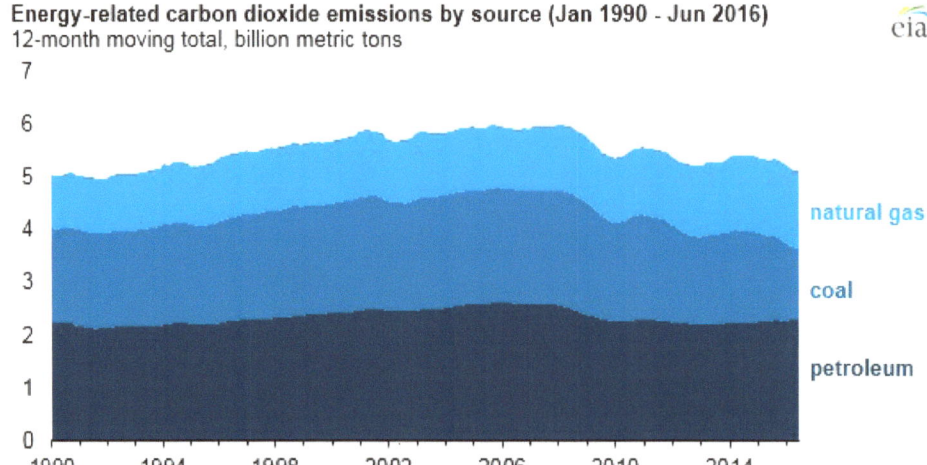

November 2016 • Issue 11

cent and 3 per cent, respectively.

Changing fossil fuel consumption mix. Coal and natural gas consumption each decreased compared to the first six months of 2015. However, the decrease was greater for coal, which generates more carbon emissions when burned than natural gas. Coal consumption fell

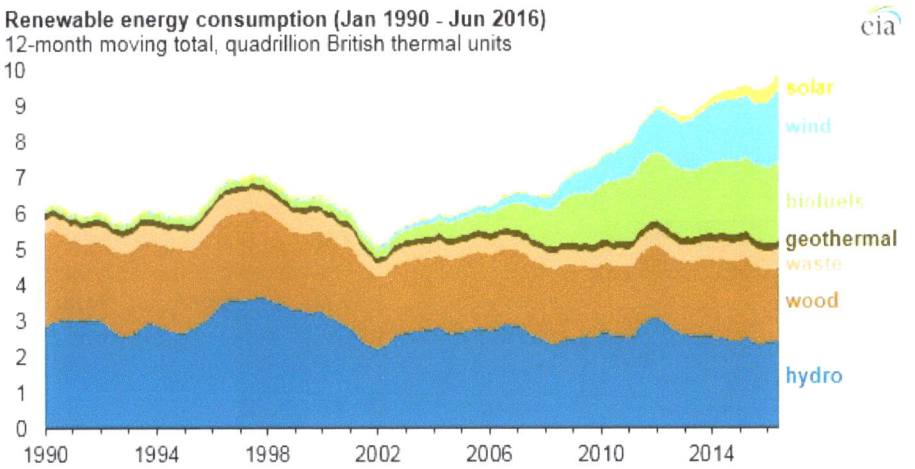

Renewable energy consumption (Jan 1990 - Jun 2016)
12-month moving total, quadrillion British thermal units

Source: U.S. Energy Information Administration, Monthly Energy Review

18 percent, while natural gas consumption fell 1 per cent. These declines more than offset a 1 per cent increase in total petroleum consumption, which rose during that period as a result of low gasoline prices.

Increasing renewable energy consumption. Consumption of renewable fuels that do not produce carbon dioxide increased 9 per cent during the first six months of 2016 compared with the same period in 2015. Wind energy, which saw the largest electricity generating capacity additions of any fuel in 2015, accounted for nearly half the increase. Hydroelectric power, which has increased with the easing of drought conditions on the West Coast, accounted for 35 percent of the increase in consumption of renewable energy. Solar energy accounted for 13 percent of the increase and is expected to see the largest capacity additions of any fuel in 2016.

An estimated 62,500 power plants are operating around the world, with a total installed generating capacity of more than 6,000 gigawatts GW in 2015. The nine largest operating power plants in the world by capacity are all hydroelectric power plants.

Four of the world's ten largest power plants are located in China, and all four of those plants began operating in the past 13 years. The world's largest dam, Three Gorges, is located on the Yangtze River and has a capacity of 22.5 GW. Hydroelectric power is the second-largest source of electricity in China, after coal, and accounted for 20 percent of the country's total generation in 2015.

South America is home to three of the world's largest power plants. Brazil's Itaipu Dam, located on the Parana River that forms the border between Brazil and Paraguay, has a capacity of 14 GW. Although the Itaipu Dam is the second-largest power plant in terms of capacity, it ranked first in the world in generation, producing 89.5 billion kilowatt-hours kWh in 2015, compared to Three Gorges's output of 87 billion kWh. Differences in seasonal flows of the Yangtze and Parana rivers account for differences in the output of the Three Gorges and Itaipu Dams, respectively.

The world's nine largest operating power plants are hydroelectric facilities

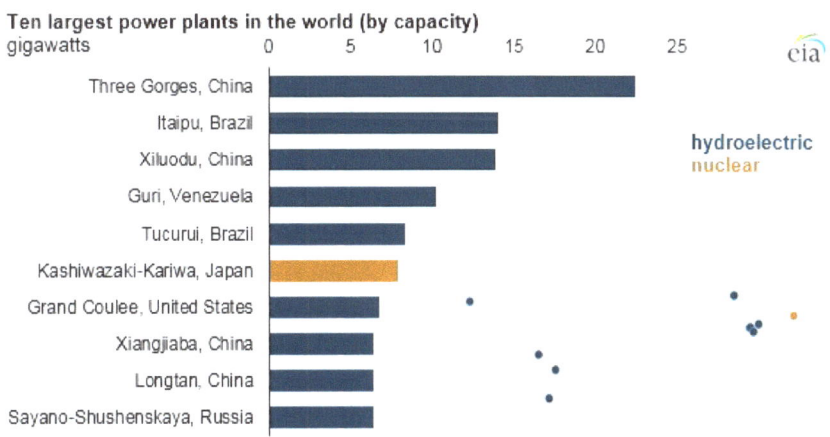

Ten largest power plants in the world (by capacity)
gigawatts

- Three Gorges, China
- Itaipu, Brazil
- Xiluodu, China
- Guri, Venezuela
- Tucurui, Brazil
- Kashiwazaki-Kariwa, Japan
- Grand Coulee, United States
- Xiangjiaba, China
- Longtan, China
- Sayano-Shushenskaya, Russia

hydroelectric
nuclear

Source: U.S. Energy Information Administration, based on International Commission on Large Dams and IAEA Power Reactor Information System. Note: Japan's Kashiwazaki-Kariwa nuclear facility has not operated since being shut down in 2011 and has not submitted a restart application.

November 2016 • Issue 11

The Kashiwazaki-Kariwa nuclear power plant in Japan is the largest nuclear plant in the world and the sixth-largest power plant of any type in the world. However, Kashiwazaki-Kariwa is among the many nuclear plants in Japan that were shut down in the aftermath of the accident at Fukushima in 2011 and has yet to file for a restart application.

The Grand Coulee Dam is the seventh-largest power plant in the world and the largest dam in the United States. It supplies power to eleven

number of turbines installed, rather than on the volume of water in the reservoir behind the dam. For instance, the Three Gorges Dam could, at the maximum level possible, hold about 10 trillion gallons of water. Dams with lower electric generation capacities, such as Venezuela's Guri and Brazil's Tucurui dams, can hold more water, with maximums of 36 trillion and 12 trillion gallons, respectively. A trillion gallons weighs about 4.2 billion tons and occupies the volume of a cube that is almost one mile or 1.56 kilometers wide.

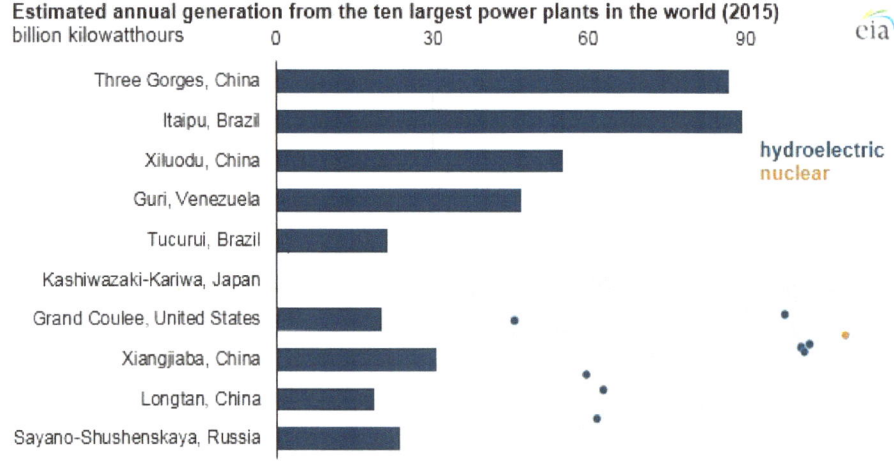

Estimated annual generation from the ten largest power plants in the world (2015)
billion kilowatthours

hydroelectric
nuclear

Source: U.S. Energy Information Administration, compiled from various sources

Several countries are building large power plants to meet growing electricity demand. Some of the soon-to-be-largest power plants in the world are hydroelectric plants under construction in countries such as China and Brazil.

states, Washington, Oregon, Idaho, Montana, California, Wyoming, Colorado, New Mexico, Nevada, Utah, and Arizona, as well as Canada. Grand Coulee Dam was the largest power plant in the world from 1949 through 1960, when power plants in Russia and Canada surpassed Grand Coulee. After an expansion, Grand Coulee was again the largest in the world from 1979 through 1986, when it was supplanted by Venezuela's Guri Dam.

The Sayano-Shushenskaya Dam in Russia is the tenth-largest power plant in the world. Located on the Yenisei River, it is the largest power plant in Russia, with a capacity of 6.4 GW. Hydroelectric power accounts for 21 percent of Russia's electricity generation.

The capacity of large hydroelectric plants is generally based on the capacity and the

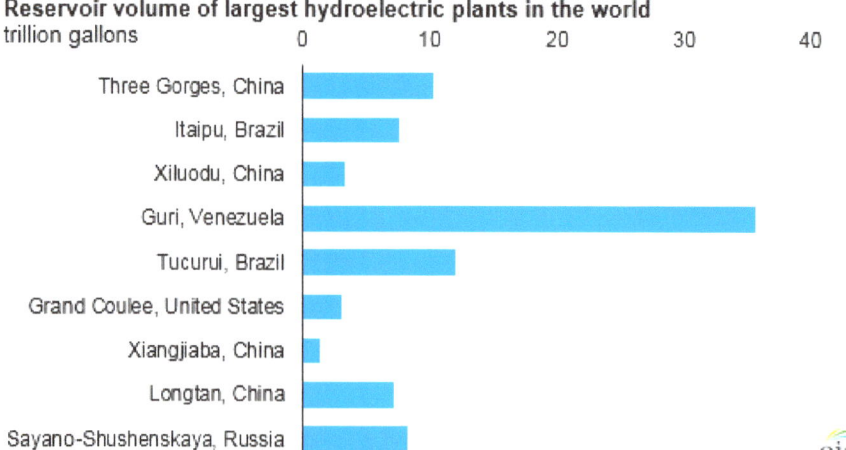

Reservoir volume of largest hydroelectric plants in the world
trillion gallons

Source: U.S. Energy Information Administration, based on International Commission on Large Dams

MECS Survey Preliminary estimates show that U.S. manufacturing energy consumption increased 3.7 percent between 2010 and 2014

Preliminary estimates show that the total U.S. manufacturing energy consumption increased about 3.7 percent between 2010 and 2014. This is the first measured 4-year increase in manufacturing energy consumption since 2002 Figure 1. Energy source shares have changed modestly since 2002, with natural gas rising from 29 percent to 33 percent

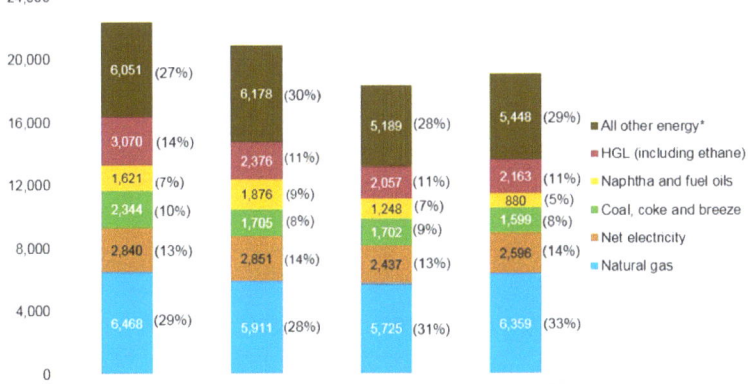

Figure 1. Manufacturing energy consumption has increased for the first time since 2002
trillion Btu

Legend:
- All other energy*
- HGL (including ethane)
- Naphtha and fuel oils
- Coal, coke and breeze
- Net electricity
- Natural gas

2002 (22,394 TBTU): 6,051 (27%), 3,070 (14%), 1,621 (7%), 2,344 (10%), 2,840 (13%), 6,468 (29%)
2006 (20,897 TBTU): 6,178 (30%), 2,376 (11%), 1,876 (9%), 1,705 (8%), 2,851 (14%), 5,911 (28%)
2010 (18,358 TBTU): 5,189 (28%), 2,057 (11%), 1,248 (7%), 1,702 (9%), 2,437 (13%), 5,725 (31%)
2014 (19,045 TBTU): 5,448 (29%), 2,163 (11%), 880 (5%), 1,599 (8%), 2,596 (14%), 6,359 (33%)

MECS survey year and total consumption of energy for all purposes

Source: U.S. Energy Information Administration
* Shipments were subtracted from all other energy

of all manufacturing energy consumption. Conversely, during this same period, the share of delivered energy to manufacturing from coal, coke and breeze declined from 10 percent to 8 percent, while the share of fuel oils and naphtha together decreased from 7 percent to 5 percent.

While manufacturing energy, consumption did increase overall from 2010, the increase was not as large as the increase in manufacturing output during that same period, implying a decrease in energy intensity. The data represent energy demand for year 2014 as reported on the U.S. Energy Information Administration's EIA Manufacturing Energy Consumption Survey MECS.

Natural gas consumption in manufacturing

increased 2 percent between 2010 and 2014. The price of natural gas continues to be low relative to many other available alternatives, and it has the lowest carbon content of the fossil fuels. In fact, several manufacturers reported renovation projects in 2014 that would permanently convert residual fuel oil equipment, a common alternative, to natural gas. They cited lower natural gas prices and preparation for expected changes in federal or state environmental regulations in their rationale to switch fuels.

While natural gas consumption has increased, coal, coke and breeze consumption in manufacturing decreased 1 percent between 2010 and 2014. Along with coal-derived coke, Figure 1 shows that this energy source continues to be phased out as less carbon-intensive energy sources, principally natural gas, become more prevalent and more price-competitive.

Manufacturing gross output rose 12 percent from 2010 to 2014. That rise outpaced the 3.7 percent increase in energy consumption, implying a decline in energy intensity. The output index represents gross output as a measure of an industry's sales, including sales to other

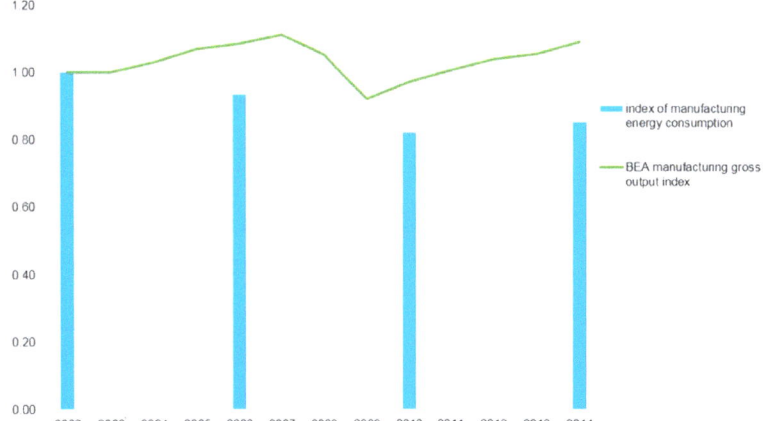

Figure 2. Manufacturing total energy consumption does not follow output levels between 2002 and 2014 suggesting energy efficiency gains
index (2002=1)

Legend:
- index of manufacturing energy consumption
- BEA manufacturing gross output index

Sources: Bureau of Economic Analysis, Gross-Domestic-Product-(GDP)-by-Industry Data (Gross Output), and U.S. Energy Information Administration

industries as well as final users. Over the long term, manufacturing energy consumption has decreased in intensity, but the relationship between gross output and energy consumption is sometimes uneven Figure 2. In fact, between 2010 and 2014, there was a 7 percent reduction in energy intensity, the ratio of a unit of energy consumption per unit of output. Once final 2014 MECS results are released, more data will be available to determine what factors and manufacturing activities are contributing to the intensity decline. For example, besides real efficiency gains, the overall manufacturing intensity decline may be attributable to a change in the proportion of high energy-intensive

industries e.g., primary metal manufacturing to lower ones e.g., textiles and machine assembly.

The data shown in Table 1 represent the first release from the 2014 Manufacturing Energy Consumption Survey. As the data are preliminary, the information may change somewhat as more detailed data tables and analysis of the 2014 MECS are released in 2017.

Table 1. 2014 MECS preliminary estimates of first use of energy for all purposes fuel and nonfuel, compared to previously available data for 2010

Energy source	2014 (preliminary)	2010 (a)	Percent change from 2010	Standard error (percent)	*= Statistically significant change
Coal	1281	1328	-3.54%	1.04%	*
Natural gas	6,359	5,725	11.07%	1.94%	*
Net electricity	2,596	2,437	6.52%	1.73%	*
Purchases	2,660	2,510	5.98%	1.73%	*
Transfers In	31	33	-6.06%	5.05%	
Onsite generation from noncombustible renewable energy	7	7	0.00%	6.89%	
Sales and transfers offsite	102	113	-9.73%	1.01%	*
Coke and breeze	318	374	-14.97%	0.61%	*
Residual fuel oil	74	170	-56.47%	0.84%	*
Distillate fuel oil	116	135	-14.07%	9.11%	
Hydrocarbon gas liquids (HGLs) including ethane	2,163	2,057	5.15%	0.59%	*
Other	6,776	6,920	-2.08%	1.27%	
Lubricants (b)	262	271	-3.32%	0.00%	-
Special naphthas (b)	69	17	308.28%	0.00%	-
Waxes (b)	15	17	-11.76%	0.00%	-
Miscellaneous nonfuel products (b)	183	159	15.09%	0.00%	-
Naphtha (c)	690	943	-26.83%	0.00%	-
Bitumen (c)	793	878	-9.68%	0.00%	-
Kerosene	39	35	11.43%	0.40%	*
Motor gasoline	6	8	-25.00%	11.54%	*
Petroleum coke	804	762	5.51%	2.11%	*
Still gas/waste gas	1,535	1,413	8.63%	0.11%	*
Pulping liquor or black liquor	810	824	-1.70%	0.49%	*
Biomass total	713	630	13.17%	5.26%	*
Agricultural waste	91	44	106.82%	1.86%	*
Wood harvested directly from trees	67	64	4.69%	4.74%	
Wood residues and byproducts from mill processing	516	504	2.38%	5.49%	
Wood-related and paper-related refuse	39	17	129.41%	78.81%	
Net steam/hot Water	555	731	-24.08%	6.25%	*
Miscellaneous	302	232	30.17%	19.03%	
Shipments of energy sources produced onsite (d)	638	788	-19.04%	0.40%	*
Total (e)	19,045	18,358	3.74%	0.93%	*

[7] Electricity-only and combined-heat-and-power (CHP) plants whose primary business is to sell electricity, or electricity and heat, to the public. Includes 0.2 quadrillion Btu of electricity net imports not shown under "Source."
Notes: Primary energy in the form that it is first accounted for in a statistical energy balance, before any transformation to secondary or tertiary forms of energy (for example, coal is used to generate electricity). Sum of components may not equal total due to independent rounding. Sources: U.S. Energy Information Administration, Monthly Energy Review (April 2016), Tables 1.3, 2.1-2.6.

U.S. Crude Oil Imports Increase During First Half Of 2016, The First Increase Since 2010

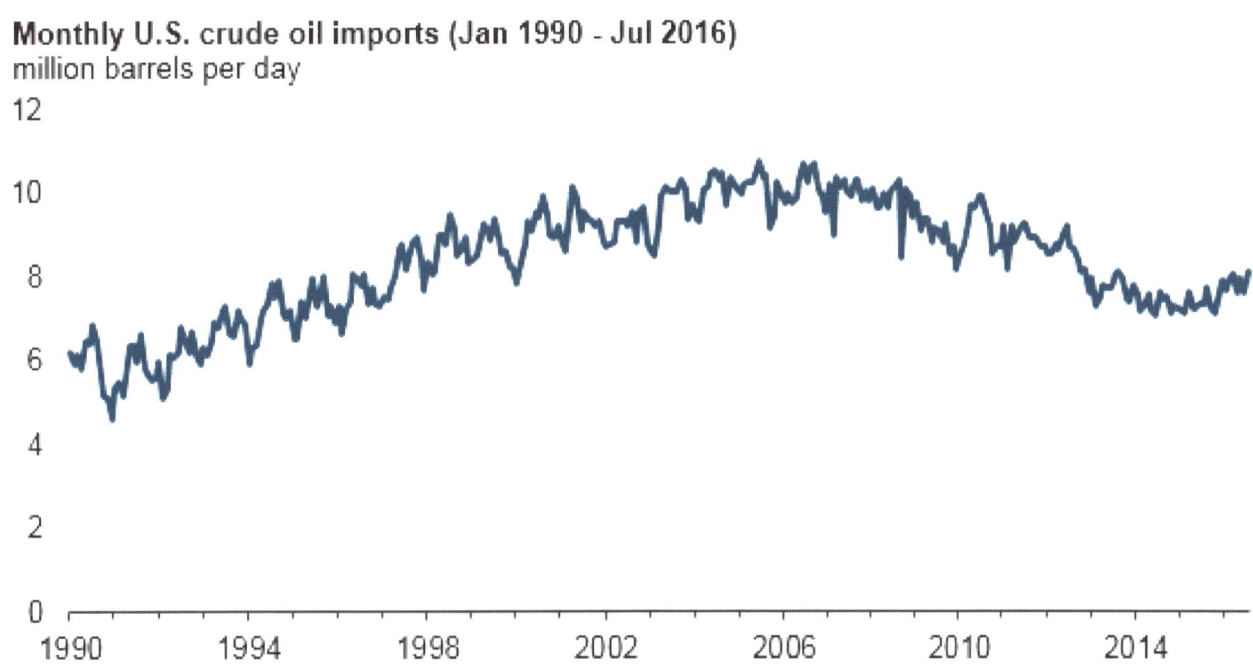

Monthly U.S. crude oil imports (Jan 1990 - Jul 2016)
million barrels per day

U.S. gross crude oil imports increased by 528,000 barrels per day b/d, or 7 per cent, during the first half of 2016 compared to the first half of 2015. This increase reverses a multiyear trend of decreasing U.S. crude oil imports as a result of increasing U.S. production.

Imports from Nigeria, Iraq, and other members of the Organization of the Petroleum Exporting Countries OPEC rose by 504,000 b/d. Declining imports from Mexico, which fell 118,000 b/d, more than offset the increase in imports from Canada, limiting the net change in imports from non-OPEC countries to an increase of less than 24,000 b/d.

Changes in crude oil price spreads were a significant

November 2016 • Issue 11

factor in the rise of U.S. oil imports during the first half of 2016. The narrowing price differences between U.S. crudes and international benchmarks provided an incentive for increased imports by refiners in areas where imported crudes now had a delivered cost advantage relative to similar domestic crudes. Additionally, lower overall crude prices contributed to a decline in U.S. crude production from an average of 9.5 million b/d in the first half of 2015 to 9.0 million b/d in the first half of 2016, resulting in higher net crude oil imports.

Canada is the largest source of crude oil imported into the United States, and its heavy crude is particularly well suited for U.S. refiners in the Midwest and Gulf Coast.

Gulf Coast PADD 3 imports increased 88,000 b/d 3 per cent, with rising imports from Middle East and African countries offsetting declines from Latin America. Imports from Iraq increased by 142,000 b/d, more than the next four countries combined. Iraq's production in 2015 rose by 700,000 b/d, enabling more of their production to be exported to the United States.

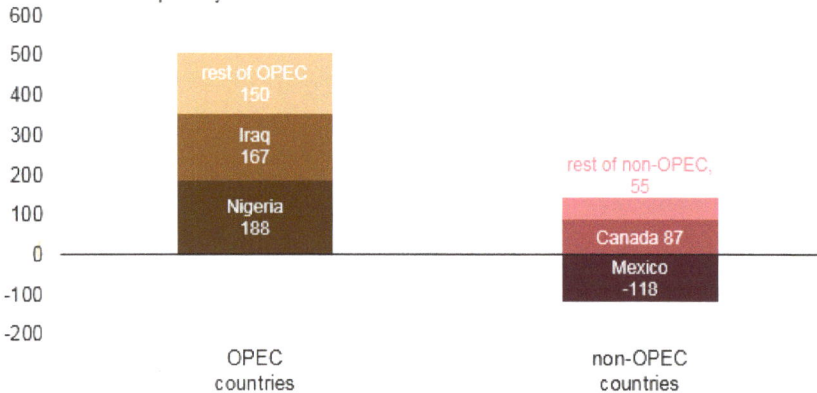

Change in U.S. crude oil imports (first half 2016 minus first half 2015)
thousand barrels per day

The Rocky Mountain region PADD 4 is the only region with declining imports during the first half of 2016, with volumes down by 24,000 b/d 9 per cent. PADD 4 is relatively isolated from import infrastructure compared with other regions, and imports have been entirely sourced from Canada for more than a decade, a trend that continued during the first half of 2016.

Imports to the West Coast PADD 5 rose by 116,000 b/d 11 per cent. Saudi Arabia, Canada, and Ecuador are the top three sources of West Coast crude imports, accounting for about two-thirds of crude oil imports into the region and about 86 per cent of the region's import growth during the first half of 2016.

As a result of shifting price, supply, and logistical dynamics, East Coast defined as Petroleum Administration for Defense District, or PADD, 1 crude imports rose by 244,000 b/d 41 per cent in the first half of 2016 compared to the same period in 2015, nearly three-quarters of which were supplied by Nigeria. Nigerian production actually declined during the first half of 2016 as a result of elevated supply disruptions. However, falling U.S. production and increasing competitiveness for seaborne light sweet crudes into the East Coast more than offset lower production levels, enabling imports from Nigeria to displace crude oil received from the Midwest PADD 2.

Although EIA's Short-Term Energy Outlook does not forecast gross crude oil imports, EIA expects annual imports of crude oil on a net basis to increase in both 2016 and 2017.

In the Midwest PADD 2, crude imports rose by 104,000 b/d 5 per cent during the first half of 2016 compared with the same time last year. Canada accounted for almost

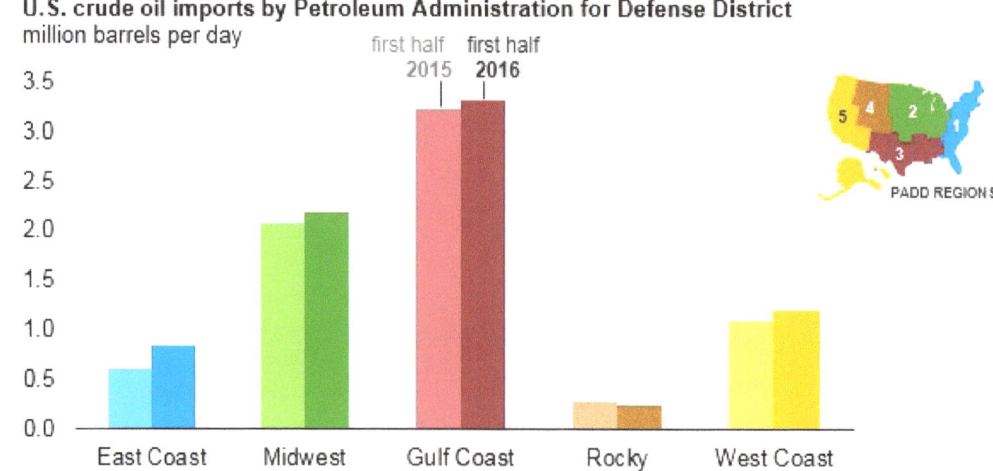

U.S. crude oil imports by Petroleum Administration for Defense District
million barrels per day

first half 2015   first half 2016

PADD REGIONS

all of the increase despite wildfires in Alberta that disrupted production later in the second quarter.

Three Gorges Dam, China Ranks First Out Of The Ten World's Largest Power Plants
Nine are Hydroelectric Facilities

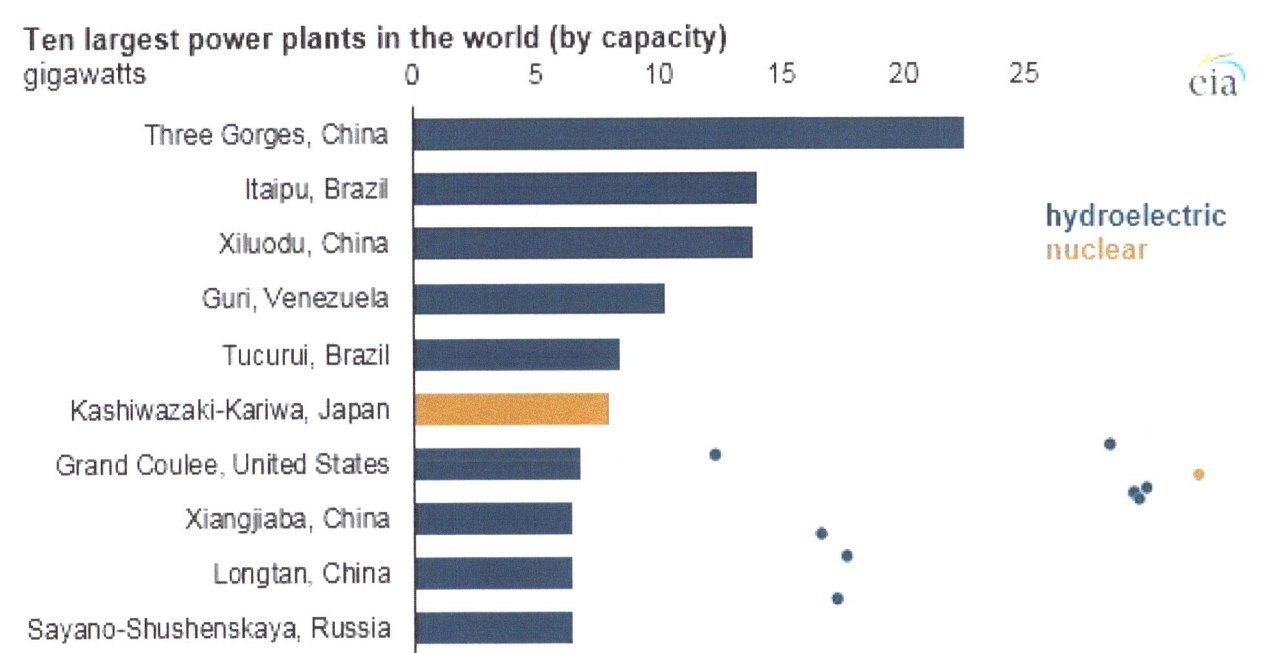

Ten largest power plants in the world (by capacity)
gigawatts

hydroelectric
nuclear

The nine largest operating power plants in the world by capacity are all hydroelectric power plants. An estimated 62,500 power plants are operating around the world, with a total installed generating capacity of more than 6,000 gigawatts GW in 2015.

Four of the world's ten largest power plants are located in China, and all four of those plants began operating in the past 13 years. The world's largest dam, Three Gorges, is located on the Yangtze River and has a capacity of 22.5 GW. Hydroelectric power is the second-largest source of electricity in China,

Contined on page 19

P O G S March 2017

The Intl Pipeline Oil and GAS Safety Conference and Exhibition

Organized by:

RGT MEDIA COMMUNICATIONS CORP.

March 14-16, 2017 Houston Texas USA

Pipeline Integrity | **Emission Reduction** | **Well Control** | **Oil and Gas Transportation** | **Chemical Extraction**

Connecting Supplier with Procurement Teams

Exhibition

200+

Exhibitors Expected

Attendance

2000+

Attendees Expected

Goal

Improve safety in the entire value chain of the oil and gas industry not limited to the well heads but distribution chains, transportation and supply chain.

Exhibit@ P O G S Safety Tech

P O G S Safety Tech provides international and local energy companies who operate across the up, mid and downstream sectors of the oil &gas supply chain with a B2B platform to meet and influence highly-focused International decision-makers and buyers.

Who is Attending?

Take Advantage of Early Registration- Register Now @

http://www.oilandgassafetyconference.com registration/online-registration/

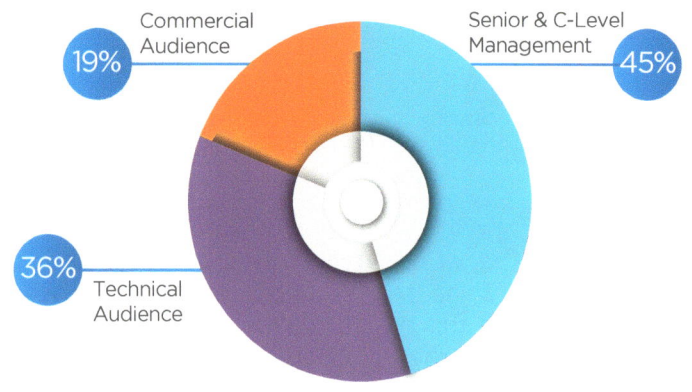

- 19% Commercial Audience
- 45% Senior & C-Level Management
- 36% Technical Audience

Who is Exhibiting?

SHOWFLOOR IS Selling Very Fast RESERVED Today

http://www.oilandgassafetyconference.com/ booth-registeration/

Official Media Partner

USA Oil and Gas Monitor
A RGT Media Communications Corp.

For further details visit website @
http://oilandgassafetyconference.com
or call +1-832-664-0618

The International Pipeline, Oil and Gas Safety Conference

POGS March 14-16, 2017   www.oilandgassafetyconference.com

Houston Astros - Minute Maid Park - 501 Crawford Street | Houston, Texas 77002

Early Discount Ends December 28, 2016

Goals

This conference seeks to address process safety issues in the upstream, midstream and downstream subsectors of the industry; with special focus on well control safety, process safety, pipeline safety, and new regulatory impact.

To help improve operational excellence in the various communities where the industry operates- emerging technologies, leak detection and prevention technologies, emission reduction technologies, compliance audit, best practices to reduce risks and hazards, and improve the overall operational safety is the focal point of this conference.

To help meet these goals - are the speakers and participating companies

Brady Austin
QHSE Service Line Owner Lloyd's Register

Mothusi Pahl
Vice President- Alphabet Energy Inc.

Vincent Higgins
Chairman and CEO Optech4D Inc

Hunter Hawa
Global EHS Director for PSRG

Robert Miller
Regulatory Compliance Specialist, Veriforce

W. Duncan Welder IV
RISC's Director of Client Services

Shoshi Kaganovsky
CEO and founder of SensoLeak

Alexis Vitone
President, AvA Excellence in Business Strategies & HSE, LLC

Tom Meek
Director of Compliance, Veriforce

Keith J. Coyle
Shareholder, Babst Calland

Mark A. Hernandez
President of Multiply Leadership

Rixio Medina
Director of Business Development for the Board of Certified Safety Professionals

Early Registration Fee - $350

Register Today for this all important industrial conference

Fill out this form email form to: *registration@oilandgassafetyconference.com*

Or mail form with check to the address below.

Mail and make check payable to: *RGT Media Communications Corp.   10777 Westheimer Street, #1100 Houston Texas 77042*

Payment Method -Card type- Amex, Visa, Master, Discovery (circle one)

Card No: _____   Expiration Date: _____   Name on card: _____   By Check Check No: _____

First Name: _____   Last Name: _____

Your Preferred Mailing Address - (Circle One) Business/ Residence

Job Title: _____   Company Name : _____   Street : _____

(No PO Boxes Please) City : _____   State: _____   Country: _____   Zip/Postal Code: ____   cut here

Day Phone: _____   Fax: _____   E-mail: _____   ✂

Program Agenda Break Down

Pipeline Safety- Leak Detection and Prevention Tech

Shoshi Kaganovsky - CEO and founder of SensoLeak

Emerging Technologies - Leveraging Virtual and Augmented Reality Technologies for Midstream & pipeline industries

Vincent Higgins - Chairman and CEO Optech4D Inc

Best Practices- Avoiding risks and hazards/ Competency-Based Training Program

Alexis Vitone- President - AvA Excellence in Business Strategies & HSE, LLC

Brady Austin - QHSE Service Line Owner- Lloyd's Register

W. Duncan Welder IV - RISC's Director of Client Services

Motivational Speaker

Mark A. Hernandez - President Multiply Leardership

Process Safety

Hunter Hawa - Global EHS Director for PSRG

PHMSA Regulations

Keith J. Coyle - Shareholder- Attorney at Law - Babst Calland

Emission Reduction Technology- Converting Flares to Power Gen

Mothusi Pahl - Vice President-Alphabet Energy Inc.

Compliance Audit- Federal/State codes and OQ NPRM

Tom Meek - Director of Compliance, Veriforce

Robert Miller - Regulatory Compliance Specialist, Veriforce

Rixio Medina - Director of Business Development for the Board of Certified Safety Professionals

---

Supporting Organization
Pennsylvania Independent Oil and Gas Association
PIOGA

Official Media Partner
USA Oil and Gas Monitor

Member Organization
Independent Petroleum Association of America
IPAA

POGS
Int'l Pipeline, Oil and Gas
Safety Conference &
Exhibition

cut here

Contined from page 15

after coal, and accounted for 20 per cent of the country's total generation in 2015.

South America is home to three of the world's largest power plants. Brazil's Itaipu Dam, located on the Parana River that forms the border between Brazil and Paraguay, has a capacity of 14 GW. Although

Estimated annual generation from the ten largest power plants in the world (2015)
billion kilowatthours

hydroelectric
nuclear

the Itaipu Dam is the second-largest power plant in terms of capacity, it ranked first in the world in generation, producing 89.5 billion kilowatt-hours kWh in 2015, compared to Three Gorges's output of 87 billion kWh. Differences in seasonal flows of the Yangtze and Parana rivers account for differences in the output of the Three Gorges and Itaipu Dams, respectively.

Four of the world's ten largest power plants are located in China, and all four of those plants began operating in the past 13 years. The world's largest dam, Three Gorges, is located on the Yangtze River and has a capacity of 22.5 GW. Hydroelectric power is the second-largest source of electricity in China, after coal, and accounted for 20 per cent of the country's total generation in 2015.

The Kashiwazaki-Kariwa nuclear power plant in Japan is the largest nuclear plant in the world and the sixth-largest power plant of any type in the world. However, Kashiwazaki-Kariwa is among the many nuclear plants in Japan that were shut down in the aftermath of the accident at Fukushima in 2011 and has yet to file for a restart application.

The Grand Coulee Dam is the seventh-largest power plant in the world and the largest dam in the United States. It supplies power to

eleven states, Washington, Oregon, Idaho, Montana, California, Wyoming, Colorado, New Mexico, Nevada, Utah, and Arizona, as well as Canada. Grand Coulee Dam was the largest power plant in the world from 1949 through 1960, when power plants in Russia and Canada surpassed Grand Coulee. After an expansion, Grand Coulee was again the largest in the world from 1979 through 1986, when it was supplanted by Venezuela's Guri Dam.

The capacity of large hydroelectric plants is generally based on the capacity and the number of turbines installed, rather than on the volume of water in the reservoir behind the dam. For instance, the Three Gorges Dam could, at the maximum level possible, hold about 10 trillion gallons of water. Dams with lower electric generation capacities, such as Venezuela's Guri and Brazil's Tucurui dams, can hold more water, with maximums of 36 trillion and 12 trillion gallons, respectively. A trillion gallons weighs about 4.2 billion tons and occupies the volume of a cube that is almost one mile or 1.56 kilometers' wide

The Sayano-Shushenskaya Dam in Russia is the tenth-largest power plant in the world. Located on the Yenisei River, it is the largest power plant in Russia, with a capacity of 6.4 GW. Hydroelectric power accounts for 21 per cent of Russia's electricity generation.

Several countries are building large power plants to meet growing electricity demand. Some of the soon-to-be-largest power plants in the world are hydroelectric plants under construction in countries such as China and Brazil.

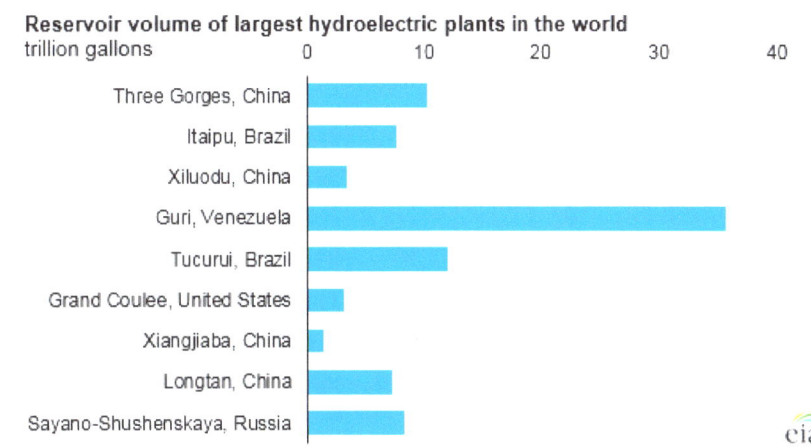

Reservoir volume of largest hydroelectric plants in the world
trillion gallons

November 2016 • Issue 11

Propane Exports Drove U.S. Petroleum Product Export Growth in First Half Of 2016

U.S. petroleum product exports (first half 2015 and first half 2016)
thousand barrels per day

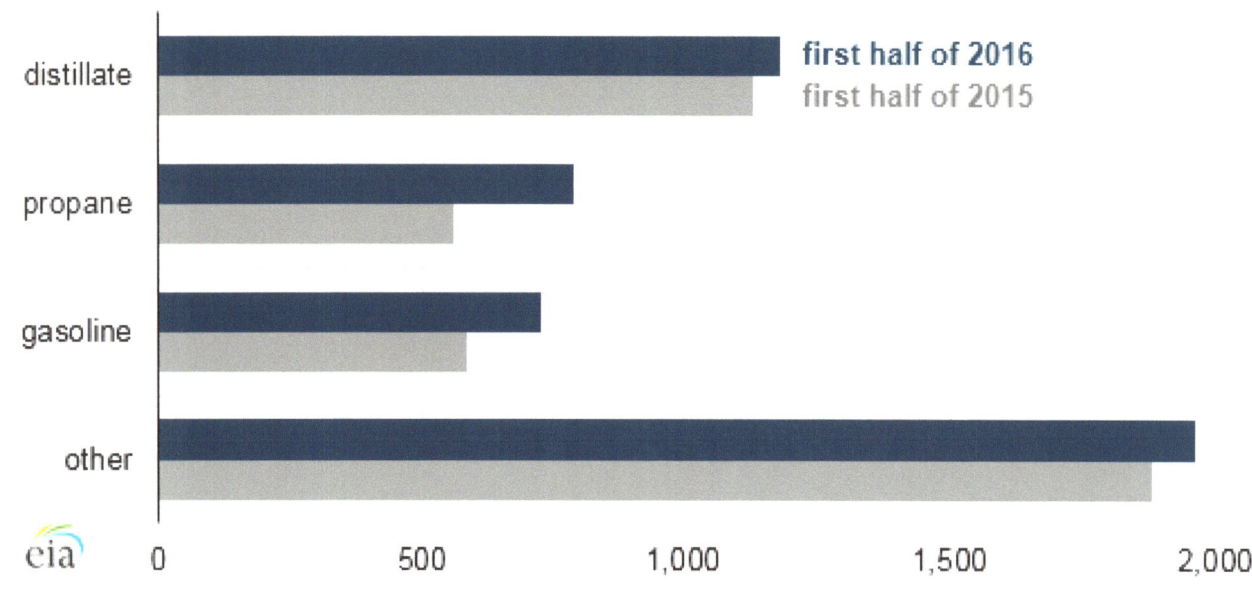

In the first half of 2016, the United States exported 4.7 million barrels per day b/d of petroleum products, an increase of 500,000 b/d over the first half of 2015 and almost 10 times the crude oil export volume. While U.S. exports of distillate and gasoline increased by 50,000 b/d and nearly 140,000 b/d, respectively, propane exports increased by more than 230,000 b/d. Propane surpassed motor gasoline to become the second-largest U.S. petroleum product export, after distillate.

Although total U.S. petroleum product exports grew, export destinations remained largely

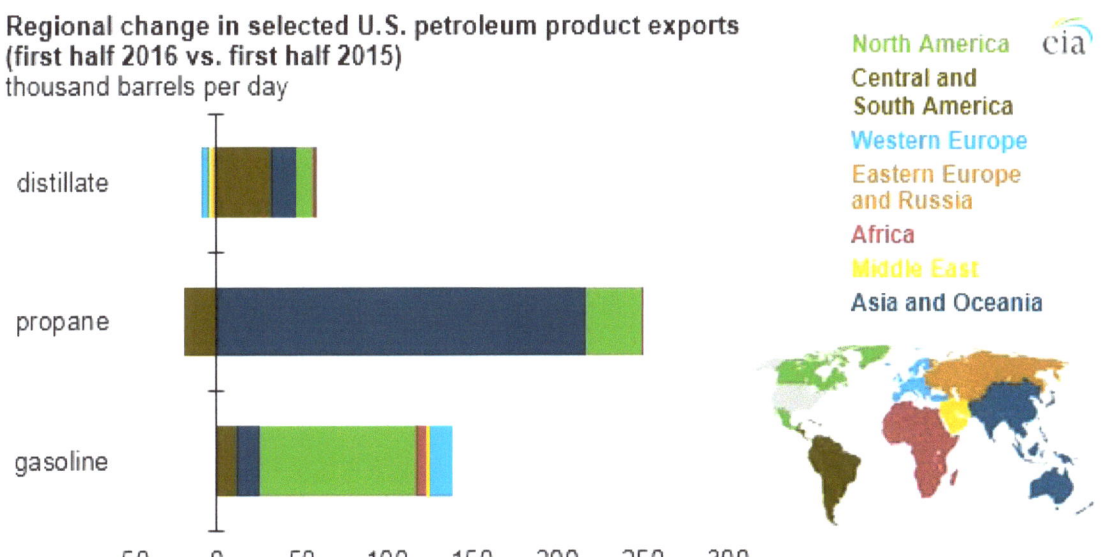

Regional change in selected U.S. petroleum product exports (first half 2016 vs. first half 2015)
thousand barrels per day

North America
Central and South America
Western Europe
Eastern Europe and Russia
Africa
Middle East
Asia and Oceania

unchanged. Mexico, Canada, and the Netherlands received the greatest volumes of U.S. petroleum products in the first half of 2016, importing 775,000 b/d, 579,000 b/d, and 271,000 b/d, respectively. U.S. petroleum products tend to stay in the Western Hemisphere. In 2015, approximately 60 per cent of total petroleum product exports remained within the Western Hemisphere, down slightly from 65 per cent in 2005.

Distillate exports averaged 1.2 million b/d in the first half of 2016, an increase of 50,000 b/d from the same period of 2015. Central and South America accounted for the largest share of U.S. distillate exports, averaging more than 620,000 b/d in the first half of 2016, up more than 30,000 b/d from the same period of 2015. The largest single destination overall for U.S. distillate exports was Mexico, which averaged 147,000 b/d in the first half of 2016.

U.S. propane exports increased from 562,000 b/d in the first half of 2015 to 793,000 b/d in the same period of 2016. Exports to Asia and Oceania accounted for 94 per cent of this growth. Japan imported the most U.S. propane at 159,000 b/d in the first half of 2016, an increase of 111,000 b/d from 48,000 b/d in the same period of 2015. U.S. exports of propane to Panama, however, fell from 41,000 b/d in the first half of 2015 to 7,000 b/d in the first half of 2016.

The **large increases in propane exports to Japan** and decreases in propane exports to Panama could be a result of reduced ship-to-ship transfer activity. Some of the propane exports from the United States that undergo ship-to-ship transfers will cite the

location of the transfer and not the final destination of the propane. This often results in larger-than-actual export numbers for the countries where the ship-to-ship transfers take place and in less-than-actual numbers for some final destinations.

Gasoline exports increased 138,000 b/d in the first half of 2016 compared with the first half of 2015. North America Canada and Mexico accounted for most of the growth, with an increase of 92,000 b/d. Similar to U.S. distillate fuel exports, Mexico represented the largest single recipient of U.S. gasoline exports at 363,000 b/d in the first half of 2016, up from 283,000 b/d in the first half of 2015.

As part of the energy reforms passed in 2013, Mexico liberalized its energy sector, allowing market participants other than the state company Petroléos Mexicanos Pemex. In January 2016, as part of the liberalization process, Mexico began to allow companies besides Pemex to import fuels, resulting in increased exports from nearby refineries along the U.S. Gulf Coast. Canada was the second-largest recipient of U.S. gasoline at 66,000 b/d in the first half of 2016, up from 55,000 b/d in the first half of 2015.

November 2016 • Issue 11

Russian Petroleum Companies and Government Revenues Declines Sharply Due to Low Oil Prices

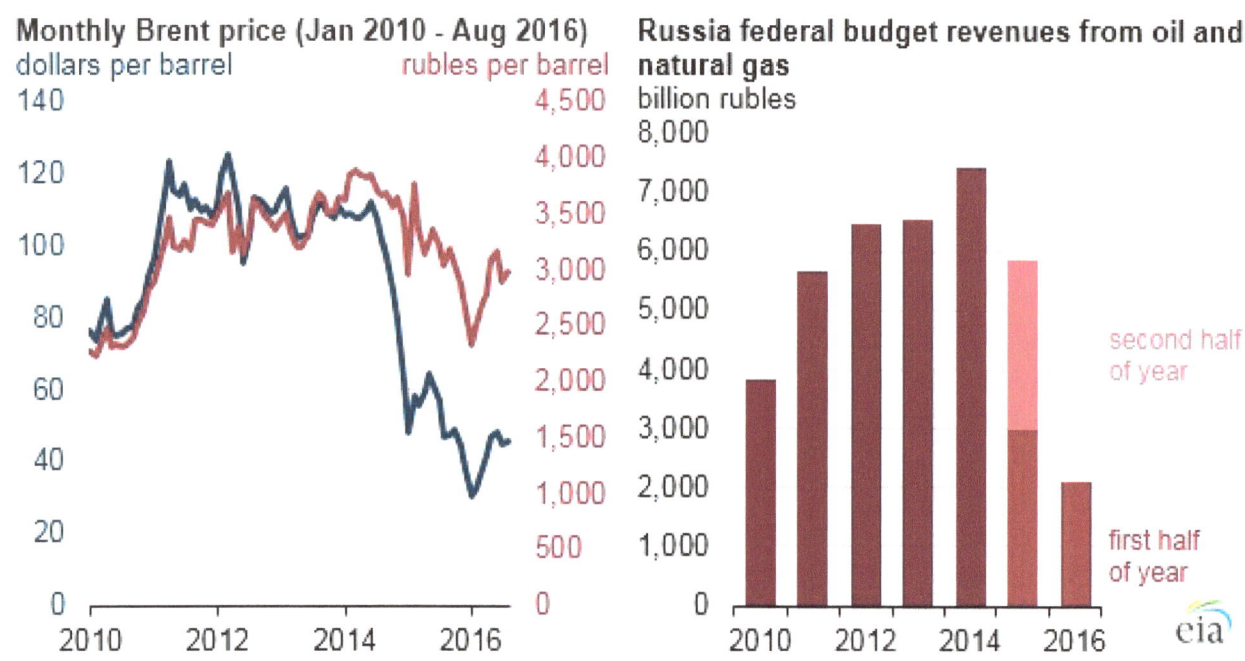

Russian federal revenue from oil and natural gas production has declined significantly in response to low oil prices. However, Russian oil and natural gas companies' capital investment programs have been less affected, if at all. **Russia's two main hydrocarbon taxes are calculated by formulas that result in lower tax rates at lower crude** oil prices. As oil prices fall, petroleum companies retain a larger share of revenue, but government revenues from oil and natural gas production fall even faster than prices.

In 2015, the Brent crude oil price, measured in U.S. dollars per barrel, declined by 47 per cent

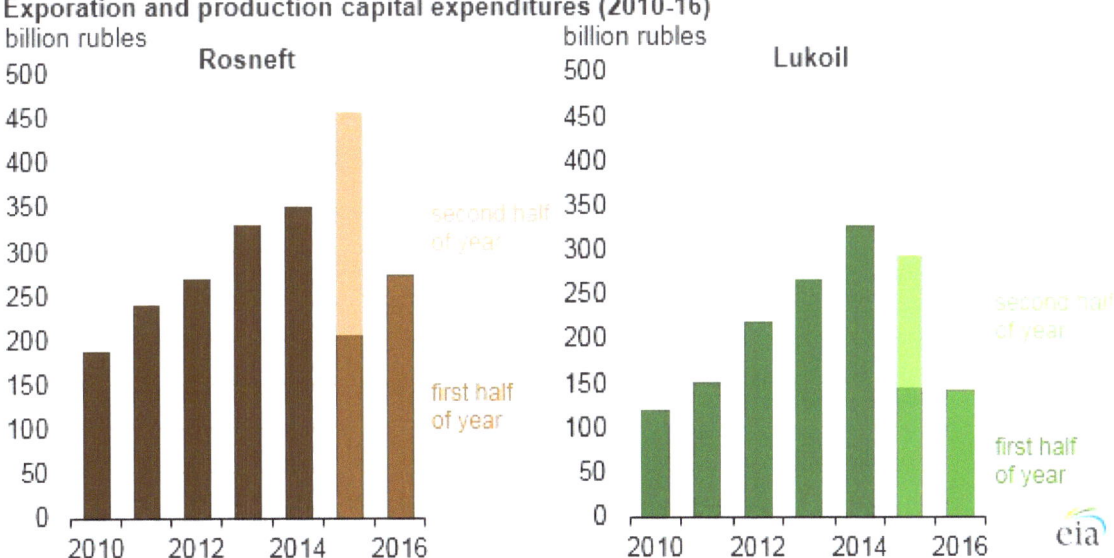

Exporation and production capital expenditures (2010-16)

versus 2014, and the price for the first half of 2016 is down an additional 31 per cent versus the first half of 2015. Because of changes in the exchange rate, the decline in the Brent price as measured in Russian rubles is not as dramatic: a 16 per cent decline from 2014 to 2015, and an additional 16 per cent from the first half of 2015 to the first half of 2016. Over the same periods, Russian federal budget revenues from oil and natural gas fell by 21 per cent and 29 per cent, respectively. Many Russian petroleum companies have, in ruble terms, increased investment or seen only modest declines in investment over the same period.

State-owned Rosneft and independent Lukoil are Russia's two largest oil producers. Together, the two companies account for about half of Russia's roughly 11 million barrels per day oil production. While oil prices and government revenues declined in 2015, Rosneft increased its capital expenditure on exploration and production for projects in Russia by 30 per cent compared with 2014. Rosneft's exploration and production capital expenditure in the first half of 2016 was 33 per cent higher than in the first half of 2015. In comparison, Lukoil's spending on exploration and production for projects in Russia declined 11 per cent in 2015 versus 2014, and expenditures in the first half of 2016 were 2 per cent lower than in the first half of 2015.

The favorable tax structure and exchange rate for Russian oil companies, the subsequent continued high investment levels at Rosneft, Lukoil, and other Russian oil and natural gas companies, and slower production declines at old fields all have helped push production to record post-Soviet levels.

Considerable investment has also been made in many new fields that have recently started up or are due to start in the near future.

The Russian government has implemented or proposed various measures to increase revenues that could affect companies' future investment plans. The Russian government has changed the two main hydrocarbon taxes, minerals extraction tax and export tax, several times in recent years. The most recent changes and proposals for upcoming changes all tend to raise the taxes paid by oil and gas companies.

In January 2015, the Russian government announced its intention to sell some of its shares in several Russian companies, including Bashneft and Rosneft. Bashneft was one of Russia's 10 largest oil producers. On October 12, 2016, the federal government sold its 50.08 per cent controlling stake in Bashneft to Rosneft, Russia's largest oil producer, for $5.3 billion. The Russian government currently owns 69.5 per cent of Rosneft. It also intends to sell up to 19.5 per cent of Rosneft, retaining a controlling interest.

In addition to taxes, **the Russian government also collects dividends from oil and gas companies in which the state is a shareholder**. In April 2016, the Russian government directed state-controlled companies to pay 50 per cent of 2015 net income out as dividends, nearly double the dividends companies would normally pay. Oil companies have objected to the tax and dividend increases, arguing they divert money from capital investment programs.

November 2016 • Issue 11

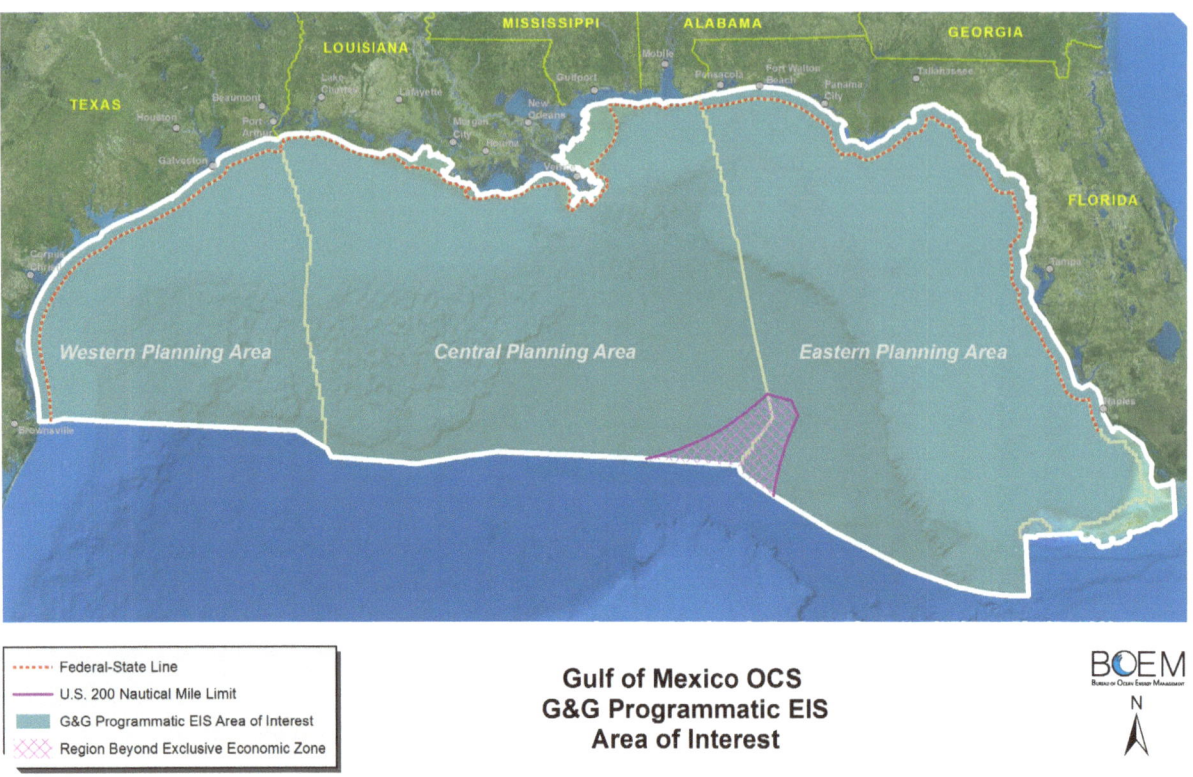

Federal-State Line
U.S. 200 Nautical Mile Limit
G&G Programmatic EIS Area of Interest
Region Beyond Exclusive Economic Zone

**Gulf of Mexico OCS
G&G Programmatic EIS
Area of Interest**

BOEM's Draft Programmatic Environmental Impact Statement for Gulf of Mexico Geological and Geophysical Surveys Announced

The Bureau of Ocean Energy Management BOEM has announced the availability of a Draft Programmatic Environmental Impact Statement PEIS that recommends strong measures to protect marine mammals and coastal environments in the Gulf of Mexico from the potential impacts of geological and geophysical G and G surveys for oil, gas and minerals.

Completion of the draft PEIS was a condition of a federal court settlement between BOEM and the Natural Resources Defense Council and other co-plaintiffs announced earlier this year.

"BOEM's approach offers the strongest practicable safeguards in an effort to eliminate or reduce impacts to marine mammals and the environment," said BOEM Director Abigail Ross Hopper. "We continue to conduct research and monitor the science of this field and work with other agencies and stakeholders to create and maintain the protection of these resources."

The draft PEIS evaluates the potential environmental impacts of G and G survey activities on marine mammals, fish, corals, and other environmentally sensitive species in the seabed and water column of the Gulf's Outer Continental Shelf. The surveys will inform oil and gas exploration and sand extraction in Federal and adjacent state waters. G and G surveys use various technologies to determine whether areas have high potential for energy development or extraction of minerals, such as sand used for coastal restoration; and to identify potential hazards and environmental concerns.

The G and G activities assessed in the draft PEIS include deep-penetration and high-resolution seismic surveys, electromagnetic surveys, magnetic surveys, gravity surveys, remote-sensing surveys and geological and geochemical sampling.

Among the mitigations BOEM has evaluated are protected species observers on each boat, mandatory vessel avoidance of marine mammals and start up/shut down rules that apply if/when marine mammals are observed in the area.

Manatee
(*Trichechus manatus*)

Whale Shark
(*Rhincodon typus*)

Bryde's Whale
(*Balaenoptera edeni*)

BOEM is the lead agency on this draft PEIS, with the Bureau of Safety and Environmental Enforcement and the National Marine Fisheries Service NMFS as cooperating agencies. The PEIS will support both BOEM's G and G permitting and NMFS's Marine Mammal Protection Act decision-making for oil- and gas-related G and G. The target date for completing the PEIS is September 2017.

The proposed project area evaluated BOEM's Western, Central, and Eastern GOM Planning Areas as well as adjacent State waters.

The G and G activities provide information about the location and extent of oil and gas resources, bottom conditions for oil and gas, and suitable locations of sand and gravel used for coastal protection and restoration. State-of-the-practice G and G data and information are required for business decisions in furtherance of exploring for and developing offshore OCS oil and gas resources, assessing sites for offshore renewable energy, and locating marine mineral resources.

API Offshore Sr. Policy Advisor Andy Radford, speaking in a press conference call, raised concern on the various mitigation measures proposed after examining the Bureau of Ocean Energy Management draft regulations.

Mr. Radford said that the restrictive nature of some of the proposed mitigations would threaten the economic and operational feasibility of performing

the number of "takes" – which is a term of art from the Marine Mammal Protection Act that includes an injury or behavioral disruption to a marine mammal – needs much more scrutiny. Historically the models used have been overly conservative and do not take into account the effective mitigation measures used by industry. The result is astronomically and unrealistically high numbers for potential takes that are then misused and sensationalized by groups opposed to energy production. The modeling used should accurately reflect the best science and research and our operational experience, which indicates that seismic surveys have little-to-no impact on marine wildlife populations.

PEIS Draft Regulations- API's Andy Radford Raise Concerns over Mitigation Measures Proposed

seismic surveys in the Gulf of Mexico. "We have been performing seismic surveys safely with no demonstrated harm to marine mammal populations for decades. Neither our operational experience nor the best available science would dictate the level of precaution proposed in certain alternatives".

Specifically concerning are the mitigation measures that would require shutting down operations for dolphins. There is nothing to justify the need for these restrictions. Also, proposed minimum separation distances for seismic surveys running concurrently are not justifiable – something that BOEM acknowledged in the Atlantic Seismic Final EIS. BOEM's analysis in that environmental review document concluded that the proposed 25-mile separation distance would only slightly reduce potential impacts to marine mammals and noted that there was "uncertainty" surrounding its effectiveness as a mitigation measure.

In addition, API is also deeply troubled by proposals to reduce the overall number of seismic surveys performed in the Gulf of Mexico by 10 and 25 percent. This is problematic for several reasons including potentially lower production in the future, but mainly the potential economic impact on the Gulf of Mexico region because of the seismic surveys that will not be performed and as a result wells that will not be drilled. These reductions would likely have a negative effect on the regional economy and our industry's ability to create jobs.

Additionally, the methods used by BOEM to calculate

Advances in seismic imaging technology and data processing over the last decade have dramatically improved the industry's ability to locate oil and natural gas offshore. And those energy sources – especially in the Gulf – can be harnessed to create hundreds of thousands of jobs, help American consumers and strengthen our national energy security.

Continuing to perform seismic surveys will produce known discoveries more efficiently and will help uncover new oil and natural gas resources in the Gulf. This will allow people to make informed decisions about the potential for continued job creation and economic growth from offshore energy production in the Gulf of Mexico.

As history, has shown time and time again, exploration and development activities generally lead to increased resource estimates. For example, in 1987 the Minerals Management Service estimated 9.57 billion barrels of recoverable oil resources in the Gulf of Mexico. With more recent seismic data acquisition and additional exploratory drilling, and technological advances that estimate rose in 2011 to 48.4 billion barrels of oil — a fivefold increase.

Additionally, a rigorous environmental analysis and permitting process ensures that seismic surveys are properly managed and conducted so they have minimal impact on the marine environment. There is no evidence that the sound produced by exploring for oil and gas with seismic surveys has resulted in any physical or auditory injury to a marine mammal or impacted marine mammal populations in the Gulf of Mexico.

In fact, marine life and commercial fishing have thrived in the Gulf of Mexico for more than 30 years.

According to BOEM's chief environmental officer Dr. William Brown, seismic surveys are frequently used in the Gulf of Mexico with no known detrimental impact to marine animal populations or to commercial fishing. In BOEM's own words, "there has been no documented scientific evidence of noise from air guns used in geological and geophysical G and G seismic activities adversely affecting marine animal populations or coastal communities."

And in addition to the oil and natural gas industry, seismic surveys are commonly used by the U.S. Geological Survey, the National Science Foundation, and the offshore wind industry.

In conclusion, the industry remains committed to improving the scientific understanding of the impacts of our operations on marine life. Seismic surveying in the Gulf of Mexico is a critical part of safe offshore energy development that is necessary if we are to continue to harness our nation's energy potential for the benefit of American energy consumers.

U S Total Coal Mine Production fell from between 2010 and 2014

U S Total Coal Mine Production fell from between 2010 and 2014 by 84,319,390 production. This is due to total decline in Production from each basin during that period time. In the meantime, production output from each coal basin shows that the western region basin produce more coal than other basin in the United States. Powder River Basin was second, Appalachia, Interior Region, Illinois Basin, Northern, Central, other western, other interior, Uinta Basin, and southern Basin, respectively.

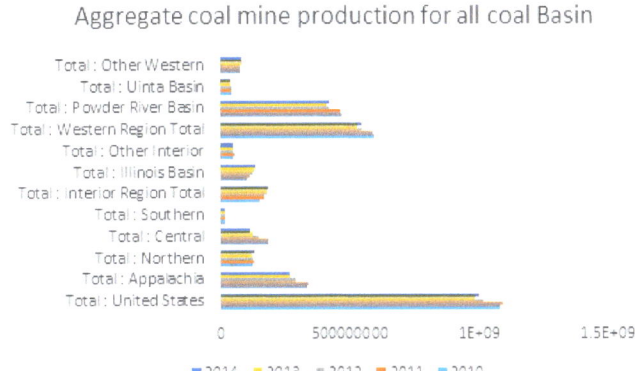

Aggregate coal mine production for all coal Basin

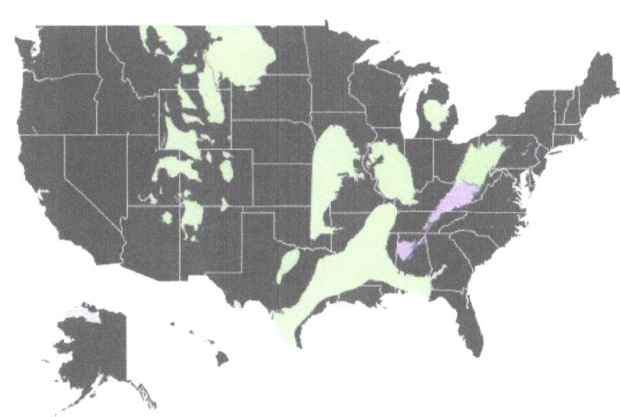

Aggregate coal mine production for all Coal Basin in the USA

description	2010	2011	2012	2013	2014
United States	1084368148	1095627536	1016458418	984841779	1000048758
Appalachia	335247719	336017185	291929342	269671505	266978782
Northern	129190963	132133051	124819209	123436761	133998444
Central	186141816	184813084	147788994	127614727	116617393
Southern	19914940	19071050	19321139	18620017	16362945
Interior Region Total	155652780	170326890	179961429	182993677	188604440
Illinois Basin	105088852	116031842	127269712	132127990	137180960
Other Interior	50563928	54295048	52691717	50865687	51423480
Western Region Total	591610917	587633022	543243909	530210328	542841927
Powder River Basin	468428341	462600212	419066016	407566885	418156236
Uinta Basin	43698593	45541810	43950154	39520353	40122274
Other Western	79483983	79491000	80227739	83123090	84563417

November 2016 • Issue 11

The Energy Transition Could Come Faster Than We Think- Spencer Dale, BP's Group Chief Economist

The energy industry faces uncertainties of daunting magnitude on many levels, says Spencer Dale, BP's Group Chief Economist, in this exclusive interview: the pace of climate change policy, the growth of renewables, the apparent demise of coal, falling energy prices, the role of natural gas in the energy mix, and the likely impact of energy efficiency on demand growth. According to Dale, "it's possible that we will see forces leading to a faster transition." But he adds that renewables, solar, wind and biomass, will still "struggles to reach 10 percent of the world's energy supply by 2035.

Spencer Dale is a relatively new kid on the energy block, who has a fresh perspective to offer on the future of the global energy sector. In October 2014, he joined BP as Group Chief Economist after a career spanning a quarter of a century in the banking industry.

For six years, between 2008 and 2014 he was on the Bank of England's Monetary Policy Committee – "all the way during the financial crisis". He chose to move from banking to energy because energy "is critical for global economic growth; it's critical for thinking about the sustainability of our planet over the next 40 or 50 years; and the chance to be involved in some of those discussions and issues and debates was too good an opportunity to turn down."

He now leads the team of analysts that produces BP's annual long-term Energy Outlook and the famous annual BP Statistical Review of World Energy, widely regarded as the "industry bible" of energy statistics. This puts him in a great position to offer insights into energy trends. In this interview, he gives his views on climate policy, the ongoing transition to a lower-carbon global energy economy, major shifts in the energy mix, and likely price trends.

One of the successes of the COP 21 climate change negotiations in Paris last year was the target to limit global warming to "well below 2°C"? How optimistic are you that humankind will be able to achieve the level of co-operation needed to achieve that target?

The global energy industry – governments, regulators, resource owners, producers like BP – faces two massive challenges over the next 20-30 years. The first is to ensure that we use energy in a sustainable way consistent with the long-term health of the

November 2016 • Issue 11

planet. At the same time, it's important to make sure there are plentiful energy supplies for the fast-growing economies of the world – so that many hundreds of millions of people can be lifted out of low incomes, out of fuel poverty.

Paris was a significant step forward in addressing the first challenge. One of the messages from both our annual Energy Outlook and the BP Statistical Review of World Energy is the scale of the change that we will need to see to get close to achieving the goals set out in Paris – in terms of both energy efficiency and the fuel mix. That will require significant changes in policy, technology and consumer behavior.

Your projections show that we will still depend heavily on fossil fuels by 2040. How likely is it that we might see disruptive change that means the transition to a lower-carbon energy economy happens much faster than people currently foresee?

It's possible that we will see forces leading to a faster transition coming from a number of different fronts – and they may all operate together. One is policy. What was striking about Paris was not only the Intended Nationally Determined Contributions INDCs that governments pledged but also the commitment to come back and review those pledges, to find further policy support in the future. Another is technology, which is likely to change very dramatically, both on the supply side and the demand side.

But – history tells us that it takes an awful long time for new energies to gain market share. This year's Statistical Review showed a chart which looked at the evolution of different fuels from the point at which they achieved 1percent of the world's energy supply. We then looked at how that share increased over the next 50 years. It took more than 40 years for oil's share to rise from 1 percent to 10 percent. Gas, even after 50 years, still didn't provide 10 percent. In our Energy Outlook, we have renewable energy – meaning wind, solar and biomass – growing more quickly than any fuel in history. It still struggles to reach 10% of the world's energy supply by 2035.

A big theme in the Statistical Review was the decline of energy prices, especially oil and gas. How do you see prices evolving?

Over the next five years we are likely to see a gradual firming in oil prices. There are three things I'm looking at closely to get a sense of what's going on. One is

> ## "I find it frustrating that so many people when talking about the Paris Agreement almost exclusively focus on the supply side"

a very significant stock overhang of oil inventories which is likely to act as a dampener on the pace at which oil prices rise. Secondly, it is striking that we've seen, for a couple of months now, the US tight oil weekly rig count rise week after week. Thirdly, working in the opposite direction, is a very significant fall in investment spending. Capex spending in oil in gas this year is likely to be around a third lower than in 2014. That has been partially offset by falls in costs but even so real investment has shrunk and that is likely to squeeze supply growth.

Those three things together will have an important bearing on how prices evolve over the medium term. Beyond that, I don't know. It's unlikely that prices will jump back to the levels we saw in 2010-2014. If you look at that period, that was a function of quite specific factors, particularly the severity of supply disruptions in the Middle East and North Africa post the Arab Spring.

"I find it frustrating that so many people when talking about the Paris Agreement almost exclusively focus on the supply side"

For gas, there's a similar excess supply that needs to work through, in terms of very strong growth in US shale gas, compounded by weakness in Asian demand, and a growth spurt in global LNG supplies. Again, I expect the market to gradually absorb that supply so we'll see some firming in gas prices. But I am struck by the supply potential – particularly in US shale gas – which is likely to limit gas price rises in the medium to longer term.

The shale gas and shale oil revolution has transformed the energy landscape in North America. Do you expect to see this revolution replicated outside North America?

We expect North America to continue to dominate the growth of tight oil and shale gas over the next 20 years. There's a perfect set of factors in America which enables this to happen. If I look around the world, no other country comes close to meeting that perfect set of factors. However, in our Energy Outlook we do project gradually emerging growth in the rest of the world, such that towards the end of our outlook, in 10-20 years' time, around half of the increase in production is coming from outside of North America.

Natural gas has become highly controversial, with some seeing it as key to mitigating climate change

November 2016 • Issue 11

and others dismissing it as just another fossil fuel – and therefore part of the problem. What's your view? In particular, how important is methane leakage?

There's a danger of people lumping all fossil fuels together. Not all fossil fuels were made equal. Until there is an economically viable solution to large-scale storage of renewable power, we will need a balancing fuel to solve the intermittency problem. So, a gradual crowding out of coal and a switch to a combination of natural gas and renewable energy in the power sector is a key part of the transition that we need to see. Methane emissions are an important issue and the industry is very attuned to that. But the big picture is that natural gas is an awful lot cleaner than coal.

Generally, there seems to be a lot more emphasis on energy supply than on how we consume energy. How

do you see world energy demand growth evolving? And how big a role will energy efficiency play in limiting growth?

I find it frustrating that so many people when talking about the Paris Agreement almost exclusively focus on the supply side. Energy efficiency needs to play at least as big a role, if not a bigger role, in responding to the challenges. In our Energy Outlook, we expect global GDP to more than double over the next 20 years, while energy demand increases by only 30%. The difference between those two things is improving energy efficiency or declining energy intensity. That is critical in underpinning the shift we expect to see in the rate of growth of carbon emissions. We think the energy intensity of GDP will decline much more rapidly than we've ever seen before.

Shell Divests Non-Core Shale Acreage in Western Canada For Total Consideration Of $1 Billion

Royal Dutch Shell plc, through its affiliate Shell Canada Energy has announced it has agreed to sell approximately 206,000 net acres of non-core oil and gas properties in Western Canada to Tourmaline Oil Corp. for a total consideration of approximately $1,037 million C$1,369 million. The

consideration is comprised of $758 million in cash and Tourmaline shares valued at $279 million. Subject to regulatory approvals the transaction is expected to close in the fourth quarter of 2016.

The acreage includes 61,000 net acres in the Gundy

area of Northeast British Columbia, Canada, and 145,000 net acres in the Deep Basin area of West Central Alberta, Canada. The assets are a combination of developed and undeveloped lands, along with related infrastructure, producing 24,850 barrels of oil equivalent per day boe/d of dry gas and liquids.

"Shell retains a significant shale position in Canada and we are actively working to mature our attractive core asset base in the Montney and Duvernay," said Andy Brown, Shell Upstream Director. "At the same time we are strengthening our shales business and creating shareholder value by selling assets that do not fit our near-term development plans."

Shell has a large shales portfolio focused on North America and Argentina, and is currently maturing this portfolio as a growth option for beyond 2020 with material value and substantial long-term potential.

In Canada, Shell retains approximately 430,000 net acres in the Duvernay liquids play in Alberta and approximately 218,000 net acres in the Montney gas play in Northeast British Columbia.

Shell also has material shale positions in the United States in the Permian and Appalachia Marcellus/Utica basins and Haynesville, and in the Vaca Muerta in Argentina.

Production from Shell's Americas shales portfolio, excluding the divested assets in this release, is approximately 250,000 boe/d.

API-Implementing EPA's Control Techniques Guidelines without proper scientific input is bad public policy

API Senior Director of Regulatory and Scientific Affairs Howard J. Feldman cautioned that EPA and states should not pile on additional guidelines and regulations on the oil and gas industry until EPA completes its Oil and Gas Information Collection Request ICR and subsequent analyses.

"Moving forward with these guidelines without robust data could impose unachievable emission reduction requirements on the industry, while adding potentially significant costs to the American economy, jobs, consumers and the environment.

"Air quality has already improved dramatically over the past two decades and will continue to improve as the industry continues to deploy innovative technologies and the EPA and states implement existing standards, which are the most stringent ever.

"The United States leads the world in both production of oil and natural gas and in the reduction of carbon emissions. We are second to no one. These environmental milestones are confirmed by EPA's own GHG inventory, which has consistently shown a downward trend in emissions over the past decade,

and by EIA data showing that carbon emissions have already reached 25-year lows due to the increased use of natural gas. These improvements are projected to continue.

"This is quantifiable support showing that increased U.S. energy production, improved air quality and environmental progress are not mutually exclusive. These trends are indicative of what our industry, when given the freedom to innovate, can achieve to improve the environment as we bolster our nation's energy security.

"In light of current and proposed state and federal regulations that address existing sources, it is better to allow completion of the ICR to inform whether CTGs are warranted and avoid the risks that acting on insufficient scientific data and conflicting guidelines could impose on the American public. If the EPA fails to follow the science, we call on Congress to avoid potential barriers to American economic and environmental progress."

API has supported the Oil and Gas ICR so that any additional regulatory decisions would be based on sound science and better informed on actual emissions and cost impacts for existing sources. API also supports providing ample time for states and businesses to meet the ozone standards, which are set close to background levels.

www.ingramcontent.com/pod-product-compliance
Lightning Source LLC
Chambersburg PA
CBHW050419180526
45159CB00005B/2336